The Positive Side of Occupational Health Psychology

Marit Christensen · Per Øystein Saksvik
Maria Karanika-Murray
Editors

The Positive Side of Occupational Health Psychology

Springer

Editors
Marit Christensen
Department of Psychology
NTNU
Trondheim
Norway

Maria Karanika-Murray
Department of Psychology
Nottingham Trent University
Nottingham
UK

Per Øystein Saksvik
Department of Psychology
NTNU
Trondheim
Norway

ISBN 978-3-319-66780-5 ISBN 978-3-319-66781-2 (eBook)
https://doi.org/10.1007/978-3-319-66781-2

Library of Congress Control Number: 2017954466

This is translated into the English language, enhanced and heavily revised from the Norwegian Arbeidshelsepsykologi på norsk, published by Fagbokforlaget in 2015, Authors: Per Øystein Saksvik & Marit Christensen. This translation has been published with the financial support of NORLA.

Printed on acid-free paper

This Springer imprint is published by Springer Nature
The registered company is Springer International Publishing AG
The registered company address is: Gewerbestrasse 11, 6330 Cham, Switzerland

Foreword

The aim of this book is to present the positive side of occupational health psychology. Research within the field of occupational health psychology has traditionally focused on investigating risk factors like job demands and stressors in the workplace and the resulting health impairment process. However, it is difficult to understand the dynamics behind the factors promoting well-being, health and meaning through studies of illness, dissatisfaction and suffering. A positive focus provides new and interesting approaches to challenges and opportunities in today's work life. Positive occupational health psychology has been influenced by positive psychology, positive organizational behaviour and positive occupational scholarship, which contribute to complement occupational health psychology. The positive occupational health movement promotes an integrated and comprehensive approach that we are discussing in several of the chapters in this book. This requires a perspective where we understand and promote positive factors at work, while continuing to investigate risk factors and occupational ill health (Bakker & Derks, 2010).

The world of work is continuously changing, and change is often described as the only stable factor in organizations. This imposes high demands on employees not only in the form of high workload and time pressure but also in the form of more psychological demands as to how to cope with the social and technical innovations. In times of insecurity regarding professional future and the growth of the global economy, the importance of occupational health issues becomes even more significant. Organizations are also increasingly more aware of how the quantity and quality of employee contributions are essential business matters. Modern organizations have therefore begun to have more focus on how to manage human capital rather than an emphasis on control and cost reduction. Positive occupational health psychology has a lot to offer regarding well-being and health of employees, and a focus on obtaining and promoting job resources can also ensure sustainable performance and productivity.

Most of the authors of this book have a background in the Scandinavian context, where they have conducted most of their research. The history of occupational health psychology highlights a key role for the Nordic countries (mainly Norway,

Finland, Sweden, Denmark and Iceland), with pioneering research on the alienation of work and the dangers of Taylorism and a broader concern for the impact of work on health and alternative ways of work organization. This can be traced back to the Nordic model of work organization (Gustavsen, 2011; Hasle & Sorensen, 2013), which emerged in the 1930s and

> "was based on the idea of reducing conflicts in working life though pooling the measures available to, respectively, the employers, the unions and the government resulting in substantial packages covering a broad range of measures and institutions. With less conflicts productivity would increase, making room for wage increases, reduction in working time, and welfare programs" (Gustavsen, 2011, pp. 466).

We would like to use the foreword to discuss positive occupational health psychology in relation to the Nordic model and to argue that it might add some new insights in addition to what we already know. We seek to demonstrate how work and health and the discipline of occupational health psychology can be understood from the more positive and collectively oriented Nordic perspective, with the aim of promoting good work and yielding healthy, stimulating and productive work environments.

Although the Nordic model and Taylorism have shared the same aim of maximizing productivity at work and rebuilding collapsed post-war economies, the two traditions were in stark contrast in terms of the process by which this could happen. The Nordic model has, at its core, "a cooperation between organized labour market parties, and between these parties on the one hand and the state/the public on the other" (Gustavsen, 2011, p. 464) and defines the elements of good work such as autonomy, variety, learning and participation in decisions. Taylorism, on the other hand, advocated specialization and rationalization of work. Alternatives have emerged, including the human relations movement, the quality of working life movement and the lean enterprise movement (Gustavsen, 2011), resulting in a substantial number of research and national initiatives that have played a crucial role in setting a course or direction in work organization. Distinguished figures have emerged, most notably from Norway (e.g., the work of Lysgård in the 1960s; Thorsrud and Emery in the 1970s; and Ursin in 1970s; see also Emery and Thorsrud (1976); Thorsrud & Emery, 1970) and Sweden (e.g., the work of Theorell, Gardell and Levi separately in the 1970s), whereas in Finland the first occupational health psychologist was appointed at Kymmene paper mill in 1974, and in Sweden we had the first scientific studies of 'alienation' of workers, monotony and participation (by Gardell in the 1970s).[1] This tradition is also reflected in the European social model which combines the aims of economic growth with the aims of good working conditions and living standards, and the 1989 European Framework Directive (89/391/EEC, https://osha.europa.eu/el/legislation/directives/the-osh-framework-directive/1) which was transposed into national law by 1992 by European Member States. A number of cultural, social and economic characteristics

[1] For a broader overview of the history of occupational health psychology, please see Schonfeld and Chang (2017), Occupational Health Psychology: Work, Stress, and Health. Springer.

have been linked to the Nordic model, underlined by a belief that individuals have the ability to take control of their own future (social constructivism; Gustavsen, 2011). With variations within the Nordic countries group, these contextual characteristics have been institutionalized as shared attitudes towards work, the responsibility of organizations towards employees and the focus on worker health and productivity. The benefits are tangible: "So far, this has placed he Nordic countries in the lead in making the idea of good notion come real. In fact, in practically all types of international comparisons—ranging from economy to health and well-being—the Nordic countries come out with high scores, and if a broad range of such studies are combined, they come out on top as a group of their own" (Gustavsen, 2011, p. 479). Correspondingly, research also reflects a widespread agreement on good work and essential principles such as autonomy, variety, ownership and participation, learning and positive work relationships.

The question then becomes whether and how we could implement these principles outside of the unique context from which they emerged. At the broader level, this tradition has counterparts in, but has not necessarily directly influenced, a range of practical mechanisms and tools. Examples are the tripartite model of social dialogue, social innovation and more specifically, workplace innovation and local examples of organizations explicitly built on the principles of employee participation and workplace democracy. Notable examples are Semco in Brazil, Sekem in Egypt, SMUD in the USA and Vaude in Germany (Bal & de Jong, 2016). Although the Nordic countries are leading on new work organization (e.g. shorter work week, participation), it is possible for organizations and perhaps other countries to apply elements of the Nordic model of work organization without going through the same process of social construction, as these example organizations demonstrate. This book is inspired by the Nordic model and built on the positive, comprehensive and collective perspective, which also underlies the movement of positive occupational health psychology.

In conclusion, there is a growing movement among researchers and practitioners in occupational health psychology to have more integrated and comprehensive approach to workplace health. This approach aims to have a balance between prevention of illness at work and promotion of well-being and health by enhancement of strengths, resources and optimal functioning in the workplace. We hope this book will contribute to build a bridge between knowledge and practice because an evidence-based approach to positive occupational health could contribute to psychologically healthy workplace practices that foster employees' health and well-being while enhancing organizational performance. This book offers some lessons aimed for researchers, organizations, employees and HR-professionals and seeks to show how work and health and the discipline of occupational health psychology can be understood from the more positive and collectively oriented perspective inspired by the Nordic model.

References

Bakker, A.B., & Derks, D. (2010). Positive occupational health psychology', In S. Leka & J. Houdmont (Eds.) *Occupational Health Psychology* pp. 194–224. Oxford: Wiley-Blackwell.

Bal, P. M., & de Jong, S. B. (2016). From human resource management to human dignity development: A dignity perspective on HRM and the role of workplace democracy. In M. Kostera & M. Pirson (Eds.), *Dignity and Organizations*. Palgrave MacMillan.

Emery, F. E., & Thorsrud, E. (1976). Democracy at work. Leiden, Martinus Nijhoff Publishers.

Gustavsen, B. (2011). The Nordic model of work organization. *Journal of the Knowledge Economy, 2*(4), 463–480.

Hasle, P., & Sørensen, O. H. (2013). Employees as individually and collectively acting subjects-Key contributions from nordic working life research. *Nordic Journal of Working Life Studies, 3*(3), 9–30.

Thorsrud, E., & Emery, F. E. (1970). Industrial democracy in Norway. Industrial Relations: *A Journal of Economy and Society, 9*(2), 187–196.

Thorsrud, E. (1970). A strategy for research and social change in industry: A report on the industrial democracy project in Norway. *Information (International Social Science Council), 9* (5), 64–90. Chicago.

Acknowledgements

We are deeply grateful for the support and feedback from colleagues when writing this book. Our special thanks go to Øystein Indergård and Karianne Kvalheim for your invaluable feedback and contributions regarding proof reading. We would also like to thank Edona Zilkiqi for working with the translation of the book. This translation has been published with the financial support of NORLA.

Contents

Contributors

Marit Christensen Department of Psychology, Norwegian University of Science and Technology (NTNU), Trondheim, Norway

Thomas Christian Espenes Department of Psychology, Norwegian University of Science and Technology (NTNU), Trondheim, Norway

Anette Fjeld Department of Psychology, Norwegian University of Science and Technology (NTNU), Trondheim, Norway

Fay Giæver Department of Psychology, Norwegian University of Science and Technology (NTNU), Trondheim, Norway

Karoline Grødal Department of Public Health and Nursing, Faculty of Medicine and Health Sciences, Norwegian University of Science and Technology, Trondheim, Norway

Siw Tone Innstrand Department of Public Health and Nursing, Faculty of Medicine and Health Sciences, Norwegian University of Science and Technology, Trondheim, Norway

Maria Karanika-Murray Department of Psychology, Nottingham Trent University, Nottingham, UK

Karianne Kvalheim Department of Psychology, Norwegian University of Science and Technology (NTNU), Trondheim, Norway

Oyeniyi Samuel Olaniyan Norwegian School of Hotel Management, University of Stavanger, Stavanger, Norway

Per Øystein Saksvik Department of Psychology, Norwegian University of Science and Technology (NTNU), Trondheim, Norway

Martin Schrøder Department of Psychology, Norwegian University of Science and Technology (NTNU), Trondheim, Norway

Chapter 1
What Is the Meaning of the Concept of Work from an Occupational Health Perspective?

Per Øystein Saksvik

To have an occupation has always been crucial for the survival of the human species. This can be defined differently depending on which era we are talking about. Before the industrial revolution, and in some societies, people thought of work with disgust and contempt (Kildal, 2005). The ancient Greeks shared this view and did not have much respect for what was then considered to be manual work. This included the preparation of what was harvested from nature and was made beneficial for humans. It was only through Adam Smith's political–economical writings in the 1700s that workers could disengage from preparing natural products and instead could bring about wealth (Kildal, 2005). This, in turn, gave working a new status, as something valuable and as an abstract quantity. The contrast with the ancient Greeks, who viewed labor as near slavery, is striking. Working for our survival, driven by our needs, was not something that distinguished us from the animals, and therefore working was not considered important in a spiritual sense. It was also seen as a disadvantage to be under the control of others, i.e., not to have independence. In our part of the world, Christianity (in particular the various forms of Protestant religion) has played an important part in the consideration of the value of work. According to Christianity, our task on earth is to prepare ourselves for the eternal afterlife, by fulfilling our duty on earth in a good and worthy manner and by accomplishing the soul's immortality. "By the sweat of your brow shall you earn your bread" from the first book of Genesis is a motto for the Norwegian labor movement. In this way, work becomes a duty, as something you have to do to earn material wealth. Work is now a necessary tool for obtaining something else, and yet it has no real value in itself. Throughout the middle ages, work had mostly negative connotations. Priests and monks would perform tasks as punishment, to obtain nourishment, and to avoid idleness. As Kildal (2005) puts it: "The word (read work)

P.Ø. Saksvik (✉)
Department of Psychology, Norwegian University of Science and Technology (NTNU), Trondheim, Norway
e-mail: per.saksvik@ntnu.no

M. Christensen et al. (eds.), *The Positive Side of Occupational Health Psychology*,
https://doi.org/10.1007/978-3-319-66781-2_1

had originally a passive and negative content, and covered a large number of conditions and activities." According to the same author, humans didn't think of work as attractive and positive until the 1500s. Then, work was viewed as "humans' purpose of engagement and activity" (Johansen, 1980, p. 396 in Kildal, 2005). Work was still primarily considered as beneficial to God, not to man.

It was primarily through the writings of the father of capitalism, Adam Smith, that work was elevated to a meaning in itself, and the economic self-interest of the individual was given its power (Dahl, 2003). Marx supported working as a means for self-actualization and was opposed to the idea of capitalism which stated that work should be truly free and attractive (Kildal, 2005). A normal opinion in Norway has been that work is synonymous with paid labor. This gives rise to confusion since work is both an abstract category and a social reality that are mixed together in "The white paper on welfare (Velferdsmeldingen) 94–95" (Heen, 2008). There are many activities that are important to the individual, which are not, strictly speaking work, in the sense of employment. Moreover, it is important that the frameworks for employment are developed in today's working conditions.

The function and role of work are in constant change, and this book aims to understand how work defines itself today, and what changes we may see in the future. As the book is based on the Norwegian workforce, the definition of work will be based on where we are as a nation today, that is, what we define as the postindustrial workforce. This concept includes a movement from a place where working life has been dominated by an industrial logic toward something else (Gershuny, 2003). What this "something else" entails, is not precise, but it can involve knowledge, service, leisure, art, and culture, or a combination of these themes that will distinguish our working life in the years to come. It is, for example, only recently that we have seen Norwegian crime fiction becoming a commodity for export, making the profession of author more lucrative and allowing multiple authors to make a living even though only a few enjoy international success.

Even though the meaning of work in past years will become something else in the society of the future, we cannot deny that working has a distinct function in our society: to secure an income. The whole of western society is built upon the notion that the individual has to contribute to the community through effort and taxes. Up to 1949, a housewife's added value was a part of the national budget in Norway. This shows that the national income and an economical active population are historically negotiated (Hagemann, 2007). In modern working life, wealth creation will also have an extended role in terms of understanding what the value of work is. Creating value has been given a long-term perspective. People, for example, tend to invest in education and qualifications to get a "pay off" at a later time and to give rise to new activities and products. The same applies to what is termed as green values. Preserving the ecosystem is not achieved by having short-term profit, but rather it happens by investing in future creations. Thus, future jobs will increasingly be found within sectors which do not immediately create monetary values, but rather are linked to long-term-based competence assets.

In the Norwegian and Nordic relations, we are known internationally for having cooperation models amongst the parties in working life (Sørensen, Hasle, &

Hesselholt, 2012). This does not mean that a traditional model of work involving employer and employee, with the function of a guiding state as a third part, will be the dominant model for the future. Even though the number of self-employed people is not high in Norway (less than 10% of the total work force), the country has a large number of small companies operating in a state of turbulence (Karlsen & Lie, 2001). These small businesses change more often than other businesses, and the future of the workers is increasingly unpredictable. This book will discuss whether this is a wanted or unwanted development from an occupational health perspective.

The idea that work has a greater function for the individual than merely securing an income and ensuring survival, has been known for a long time. The classic study of Jahoda between 1930 and 1950 concluded that work, in addition to its manifest function, in terms of wages and benefits also has the latent function of providing structure in life, social contact, status, and meaning (Newell & Newell, 2002). The fact that work has a meaning beyond the creation of economic wealth was also an important point for Thorsrud, one of the major working life research pioneers on Norwegian working life (Emery & Thorsrud, 1976). He is especially known for having included work in the democratization efforts in Norway after WW2. He thought that working contributed to fulfilling a larger sum of needs, which included meaning and the development of society. These needs were on an individual level, but the idea was that they would contribute beyond the individual. Interestingly, in this context, is not that work contributes to something beyond income and wealth for the individual worker, but that these values are not only related to the economy, but also to psychological values on an individual level and higher.

Today's view of the concept of the value of work is, therefore, much more differentiated. The thought that working gives life meaning becomes a natural consequence of the historical precursors. From around the time of the new millennium, the meaning of work has assumed an important place in Danish working life research (see for example Ravn, 2008). The concept becomes differentiated and associated with individual, collective, and social meanings, but with the individual as a point of departure. Seen from this perspective, this is not all that different from the view of Thorsrud, only that "need" has switched places with "meaning." The idea that working in many ways changed identity and signification is important when talking about occupational work psychology. Now, working plays a different role, and the concept of work becomes more fluid. It is a fact that about 10% of the inhabitants of Norway get a disability benefit, when, based on their age, they should really be working. It is also true that some citizens retire before the age of retirement. In spite of all the efforts put in place for people to work beyond their retirement age, the value of not having a job has to be recognized. Many of those without a job, still work voluntarily in organizations that are dependent on these voluntary workers. Those taking early retirement could also be grandparents and this could lessen the effects of the working-home conflict (see Chap. 6). The fact that a great many people have relatively high education contributes to the changes in working life. A number of people choose studies that do not necessarily end up in a defined occupation or profession. Although many still end up in the established workforce, they bring with them values and skills that can contribute to a better understanding of the profession they eventually chose, and that will raise their own and others' self-awareness and sense of meaning. In an example from our field, the M.Sc. in Organizational Psychology, it may be that the majority of those we educate want everyday work with health, a good working environment, and safety. However, in any future job, most people will benefit from the expertise acquired from the courses we provide. In a project about temporary work, we found that a consequence of an increasing number of temporary employees is that the standard of occupational health and safety goes down (Eiken & Saksvik, 2009). If educational programs such as ours strengthen these qualifications, this will no longer be an issue with temporary employment.

One of the best illustrations of the modern working world is Charles Chaplin's movie, "Modern Times" from 1937. In the movie, we follow a little man who is a small part of a much bigger machine. The movie can be viewed as a critique of Scientific Management, which still had an impact on industry. The message of the movie is also valid in today's modern workforce where achievement still is a key concept. Sometimes, human labor is presented as something that is needed, but still something that we could do without. The production line should not be so lean that all unnecessarily slack is out of the question (Skorstad & Ramsdal, 2009). In their book, Skorstad and Ramsdal (2009) implement a critical analysis of the notion that lean production is the solution to problems and, at the same time, it gives the workers more developing working tasks. They claim that it is rather the opposite, and that the costs of these alterations are transferred to the employees who experience higher stress, bullying, and disempowerment. Nevertheless, a feature of the development of automation and innovation at work can be a decrease or a change in the number of people needed for jobs (Bain et al., 2007). This gives rise to fewer routine-based tasks and decreased manual work, which entails a transition to a highly skilled labor force, but also, maybe involuntarily as a consequence, that work must be created in new areas to maintain employment. This also increases the gap between rich and poor countries. In poorer countries, created wealth is still associated with commodities and their processing, while in the richer parts of the world there is more talk of expertise and increased technology. It is perhaps a paradox that Norway is so dependent on natural recourses such as oil, and some people claim that this limits the development of new employment in Norway.

These days we face a value debate on work. This concerns which values are brought about by work, as well as what value working has for the individual. In both areas, a lot is happening. Some years ago, only a few people thought there could be a value in communications technology, like the Internet and cell phones. Now, we see a great deal of occupations and wealth creation all around the world in association with this technology, and the trend is moving increasingly faster. What work is and what it means for the individual is also in rapid development. For many, working is no longer just a restricted activity. Life becomes about working. Naturally, this influences the consequences of the understanding of occupational health. It is perhaps this major differentiation that becomes the most problematic. At one end of the scale, we have those with marginal contact with working life: the unemployed, the recipients of disability benefit, some of those in temporarily jobs, and others who have difficulties with getting and keeping jobs (and in periods do not want to work, who in Norway survive on the social security granted by the state). Tanum interviewed unemployed subjects, who were considering working. For several, it was not a given that they would choose an occupation. They could live their lives without engaging in the workforce (Tanum, 2013). This phenomenon is still new and has not yet been examined all that much, and the long-term effects have not been researched. It could be that the need for social contact and developing an identity is now filled by social media and that blogging can be a form of creative expression that also gives meaning to people's lives.

The other extreme in relation to work also requires further investigation. The attitude can be illustrated with the concept of "Always present," used as a slogan in occupational groups, for example, real estate agents. This means that one is almost never off work. You could get questions or be approached at all hours and in all situations. A large number of occupational groups have the freedom to regulate working hours and effort, such as musicians, artists, or professional athletes. Even those relatively regulated occupations, like doctors, can have life involving working situations. These occupations have what we call boundless careers (Bertelsen, 2011; Sullivan & Arthur, 2006). Boundless refers to both psychological and physical mobility, implying that it is difficult to categorize this kind of work as a threat or as something that gives life value and content. Research done on doctors shows, for example, that they experience circumstances that make them vulnerable to presenteeism (Løvseth & Saksvik, 2014; Thun, Fridner, Minucci, & Løvseth, 2014). Presenteeism as a phenomenon will be discussed later on in this book (see Chap. 11).

This book is about the relationship between occupation and health, and when the work has changed as much as it has, it should indicate that occupational health has changed equally. Bain et al. (2007), after a thorough analysis of nine case studies, concluded that: "Work today is the same mixture of satisfaction and unpleasantness as it has always been, but the contemporary workplace is perhaps a more unstable environment that we have been used to for some time (p. 1)." Based on these researchers, what has changed in today's working life is instability and uncertainty. The implications of this are numerous, amongst others, the need to understand why and how change happens. Furthermore, is it possible to initiate measures/interventions that improve work and health, and generally, how can we develop a greater sense of security in the workforce? However, this does not mean that feeling secure is a basic premise for positive occupational health. Rather, that there are some changes in relation to security in that today's workers are more concerned with competence—or what can be labeled professional security than security based on a current position (Danielsen, Nordvik, & Saksvik, 2011).

This introduction to work has shown that work had diversity and variety. This affects the picture of how work is today, and how it likely is to develop in the future. This means that we cannot use the same set of norms for a cook employed at a hotel as for a freelance journalist. This also implies different rules for marginal workers, actively choosing to live on the state through social security, than for those working in boundless careers by choice. Freedom, boundlessness, and diversity are of course dependent on economical conjunctions and a functioning welfare state. In our introductory book on occupational and organizational psychology, we wrote that a future model for working life may be that when one ends education and until the age of retirement one can receive a fixed salary, a citizen wage (Danielsen et al., 2011). Those who, for different reasons are not capable of working, could potentially live off these wages, while others could choose to work as much as they would like (of course within some boundaries, defined by, for example, contract or the government). The model is not that far from today's reality, but has some difficulties regarding politics and wealth variables. The model can lead to a larger gap and greater inequality in society, but it addresses the issue of poverty. Some might fear the possibility that not enough people will choose to remain employed, and especially not chose what is considered hard and demanding work or low-status work. The same debate is applicable to pension reforms. Our thinking in writing this was that the future occupations all need to be upgraded in terms of becoming more stimulating and motivating, and that easy routine job will become more automated. The future will tell if this is about to become a reality. Throughout this book, we frequently discuss the theme of the development of the concept of work.

Ideas on how work should be in the future, and how this is important for the occupational health, have received a lot media attention from a Danish institution (Max Planch Odense Center on the Biodemography on Aging) (http://www.sdu.dk/en/om_sdu/institutter_centre/maxo). They claim that, in a week, 25 h should be the optimal working hours and that we should not retire until the age of 80. The reason given for that is that a shorter work day benefits all ages. For example, parents would have more time with their children and would exercise more. From a preventable perspective is it claimed that the elderly would have a job to go to (see also Chap. 10). The debate about the length of the workday is not a new one. For example, there exists an international movement that tries to implement a six-hour workday (http://www.6hourday.org). This has not received huge support amongst the politicians in Norway, but still there are examples from Norway of successes in this area (Kico & Saksvik, 2015). These successes can also be found in Sweden http://www.huffingtonpost.com/2014/06/05/sweden-working-hous_n_5446579.html), and the debate is still going on in Norway. What is optimal from an occupational health psychology perspective would be to reduce working hours for all throughout a career and aim to prolong professional life. However, it is not occupational health that governs these developments. Political constraints have a greater significance. We will examine in detail some aspects of occupational health psychology that will make us revise our views on work.

References

Bain, P. M., Baldry, C. J., Bunzel, D., Gall, G., Gilbert, K., Hyman, J. D., et al. (2007). *The meaning of work in the new economy. Future of work.* Basingstoke, United Kingdom: Palgrave.

Bertelsen, T. (2011). *En grenseløs karriere?: En studie av enkelte arbeidstakeres karriereendringer innen norske kunnskapsintensive sektorer* [The boundaryless career?: A study of some employees career changes in Norwegian expert-knowledge sectors] (Master Thesis). University of Oslo, Oslo.

Dahl, T. (2003). *Hvilken moral for dagens marked og miljø?* [What kind of moral for to-days market and environment]. In H. Brattebø (Ed.), Program for industral ecology. Report no: 3/2003.

Danielsen I. H, Nordvik, H., & Saksvik, P. Ø. (2011). Individuell karriere og lojalitet til bedriften—motstridende interesser? [Individual career and loyalty to the enterprice—contradictory interests?]. In P. Ø. Saksvik (Ed.), *Arbeids- og organisasjonspsykologi* [Work. And Organizational Psychology]. Oslo: Cappelen Akademisk Forlag.

Eiken, T., & Saksvik, P. Ø. (2009). Temporary employment as a risk factor for occupational stress and health. *Policy and Practice in Health and Safety, 07,* 75–93.

Emery, F. E., & Thorsrud, E. (1976). *Democracy at work.* Leiden: Martinus Nijhoff Publishers.

Gershuny J. I. (2003). *Changing times: Work and leisure in postindustrial society.* Oxford: Oxford university press.

Hagemann, G. (2007). *Reciprocity and redistribution: Work and welfare reconsidered.* Edizioni Plus: Pisa University Press.

Heen, H. (2008). Om bruken av begrepet arbeid [On the use of the concept work]. *Sosiologi i dag, 38,* 29–50.

Karlsen, J. E., & Lie, T. (2001). *Et Nordisk grep på arbeidsmiljøregulering* [A Nordic grip on work environment regulation]? Nordic Council of Ministers.

Kico, E., & Saksvik. P. Ø. (2015). Derailed, but implemented. In M. Karanika-Murray & C. Biron (Eds.), *Derailed organizational stress and well-being interventions: Confessions of failure and solutions for success.* Netherlands: Springer Education.

Kildal, N. (2005). *Fra arbeidsbegrepets historie: Aristoteles til Marx* [From det history of the work consept. Aristoteles to Marx]. Stein Rokkan senter for flerfaglige samfunnsstudier. Bergen: Universitetsforskning Bergen. Notat 3.

Løvseth, L., & Saksvik, P. Ø. (2014). *Prevalence, determinants and interventions to prevent sickness presenteeism among physicians in Academic Medicine. Preliminary results based on the HOUPE study and Nordic NOS-HS study.* Paper presentation at the 11th EAOHP Conference, London.

Newell, S., & Newell, S. (2002). *Creating the healthy organization: Well-being, diversity and ethics at work. Psychology at work.* Cengage Learning EMEA: Boston.

Ravn, I. (2008). Mening i arbejdslivet – definition og konceptualisering [Meaning at work—definition and conceptualization]. *Tidsskrift for Arbejdsliv, 10*(4), 59–75.

Skorstad E. J., & Ramsdal, H. (Eds.) (2009). *Flexible organizations and the new working life: A European perspective.* Farnham: Ashgate.

Sullivan, S., & Arthur, M. B. (2006). The evolution of the boundaryless career concept: Examining physical and psychological mobility. *Journal of Vocational Behavior, 1*(69), 19–29.

Sørensen, O. H., Hasle P., Hesselholt, R. R., & Herbøl. K. (2012). *Nordiske Perspektiver på Arbejdsmiljø - Mening, Indflydelse og samarbejde* [Nordic work environment perspectives—mening, control and cooperation]. NMR-rapport, Det Nationale Forskningscenter for Arbejdsmiljø. Nordisk Ministerråds rapportserie.

Tanum, M. S. B. (2013). *Arbeid? Ikke for enhver pris: En narrativ studie av arbeidslediges hverdag* [Employment? Not for any cost: A narrative study of the ordinary day of unemployed people] (Unpublished master's thesis). Universitetet i Oslo, Oslo.

Thun, S., Fridner, A., Minucci, D., & Løvseth, L. T. (2014). Sickness present with signs of burnout: The relationship between burnout and sickness presenteeism among university hospital physicians in four European countries. *Scandinavian Psychologist, 1.* doi:10.15714/scandpsychol.1.e5.

Chapter 2
What Is Health from an Occupational Health Perspective?

Marit Christensen

In understanding and explaining what the phenomenon of occupational health revolves around, we have to take a closer look at the concept of health, as well as asking the question: what is health? Answering this question is almost impossible because health is a highly complicated, complex, and context-dependent concept. Most of us agree that this concept is much more than simply the absence of illness, yet health is also dependent on culture and experience since health is something that changes throughout the entire life course. If it is so difficult to define health, then we could be asking what is the point of writing this chapter? The aim is to help the readers to be attentive to the complexity of the concept of health so that it can be used to reflect on various issues in terms of the challenges facing occupational health.

In this chapter, we first introduce a short historic view on the concept of health, starting from the early philosophers. We then focus on developments in the field of health from concern with infectious diseases to the focus on lifestyle diseases, that is, a shift from a biomedical explanatory model to a biopsychosocial explanatory model. Finally, we will discuss the differences and overlap with regard to different approaches to a health-promoting perspective. In addition, the focus will be on the health concept and how we should understand it. Here, we will make an attempt to reflect on how we define, operationalize, and measure health.

2.1 Challenges

Health issues create major challenges and costs for society. At the same time, they cause major losses for employers in terms of sickness absenteeism, sickness presenteeism, and production losses. The state of Norway annually spends 37.7 billion

M. Christensen (✉)
Department of Psychology, Norwegian University of Science and Technology (NTNU), Trondheim, Norway
e-mail: marit.christensen@ntnu.no

M. Christensen et al. (eds.), *The Positive Side of Occupational Health Psychology*,
https://doi.org/10.1007/978-3-319-66781-2_2

kroner on illness benefit, that is, about 14,500 Norwegian kroner for every employee in Norway. If all the losses are taken in account, the cost was estimated to be around 61.4 billion kroners in 2013, or 23,600 Norwegian kroner for every employee. In addition, the employer's losses have to be considered. In a report on the cost to industries (Hem, 2011), it was estimated that employers pay 13,000 Norwegian kroner every week (Hem, 2011). As an agenda in the workforce health issues should not be seen as a burden, since they produce increased opportunities and advantages. These can be in the form of reduced costs, reduced risk, and lower absenteeism and turnover. For employers, health issues can contribute to them establishing a good reputation for protecting their employees, which is favorable when collaborating with partners and for future employees. Last, but absolutely not least, placing health issues on the agenda could lead to higher productivity if the employees are healthier, happier, and more motivated.

2.2 History

The question of what health really is has been on people's minds since the dawn of time. The early Greek philosophers were later convinced that maintaining good health and fighting off disease was related to natural rather than supernatural factors. These included both physical and social health determinants and defined health as a state of dynamic balance between inner and outer surroundings. Thus, they oriented medicine toward a more scientific and humanistic perspective. The Greek philosopher Plato (428–348 B.C.) said that when looking for the reasons and treatment of disease, we should look at the human as a whole rather than focusing only on parts of the body.

Hippocrates (ca. 460–377 B.C.) is viewed as the western father of medicine and is thought to be one of the most influential people in the history of medicine. He rejected the supernatural and set the foundation for medicine as a science. Illness was not necessarily a result of the wrath of the gods, rather it was caused by environmental factors, diet, and lifestyle.

Aristotle (384–322 B.C.) believed that one should be careful when dealing with anything one could not measure. The physical is easily measurable, but the mind is not so easily measured. Based on this view, we have numerous important studies on diseases and their treatment, but the understanding is that a disease can be explained as a lack in the human's physical apparatus. Aristotle introduced eudaimonia as the highest of all human needs, that is, self-realization and the realization of one's potential.

The lack of concurrence between the philosophical thought and how to live the good life, and how human health is constructed, probably stems from the scientific revolution, where health was related to biological science. This conceptualized health in relation to the human body. Descartes (1596–1650) introduced the mind and body dualism. This means that humans consist of two separate parts: the body and the mind. Medical technology escalated this trend. Descartes suggested that the body and mind do connect at one point, "pineal gland" (later called epiphysis), but that they did not affect each other. The biomedical model is based on such thoughts.

The biomedical model of health focuses only on biological factors and excludes psychological, environmental, and social explanations. This model explains health as being the absence of disease or illness. Its main focus is on physical processes that influence our health, such as biochemical factors, physiology, or pathology. The assumption is that disease has an underlying cause; if the cause is removed, one will be healthy again (Engel, 1977; Espnes & Smedslund, 2009). This health model is closely attached to a disease model that is often described as being reductionist. This is because of the claim that it reduces disease to smaller body parts or limiting biological processes. Reductionism has become the big success of biomedical medicine. Development in this field is excessive, and its detailed knowledge is becoming even more extensive. The biomedical model emerged in the mid-nineteenth century and focuses on the physical body when diagnosing and initiating treatment which, at that time, was closely related to the challenges of infectious disease such as influenza and tuberculosis. The causes of these diseases were identified and the treatment implemented. The big breakthrough came with the discovery of penicillin. Large vaccination programs were carried out, and this led to a massive decline in the number of cases of such diseases (Engel, 1977). Based on this, the popularity of the biomedical model is understandable.

During the 1950s, there was an explosion of what were commonly referred to as lifestyle diseases (diabetes, cardiovascular diseases, cancer, etc.). These diseases compensate for the earlier infectious diseases as the main cause of death in western countries. We have also witnessed an increase in psychosomatic diseases, which include stress, burnout, back, and neck problems. Based on this development, it would be correct to assume that the existing method of medical treatment using the biomedical model is no longer adequate, and there has been a growing need for a greater focus on psychological and social processes in addition to the biomedical perspective.

The biopsychosocial model was launched as a reaction to biological reductionism and tried to build a bridge between the physical–biological and the psychological aspects of health. George L. Engel introduced in the 1970s the biopsychosocial model (Engel, 1977). Engel acknowledged the enormous contribution to the understanding of disease, but he disagreed with the view that diseases only had physical causes. He emphasized the importance of the psychological and social aspects of human illness (Matrazzo, 1980). Using the biopsychosocial model of health, Engel proposed that a more holistic understanding of the patient's biological, social, psychological, and behavioral background should be considered when reaching a diagnosis and identifying treatment. The biopsychosocial model assumes that an individual's own sense of how they feel will have great consequences for their health (Suls & Rothman, 2004) (see Fig. 2.1).

Biological influences continue to play a key role in the biopsychosocial model, where the physiological cause is identified and examined. However, biological factors are only one component of the whole picture of an illness. Psychological influences are viewed as another of the main components that can contribute to

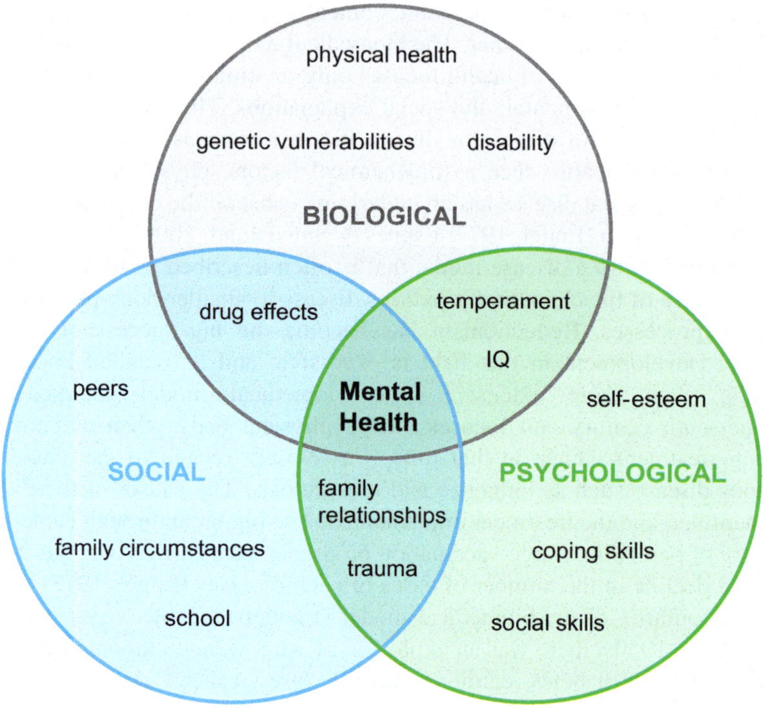

Fig. 2.1 Biopsychosocial health model

understanding the picture of an illness. This could, for example, include depression, health behavior, and negative thought patterns. The third key component is sociological influences, where we examine illness from a sociological perspective. Here we include, for example, economical relations, religion, and technology as important contributors to the development of disease. This model is based on the treatment of illness but it also includes health promotion as an essential component due to its focus on how to maintain and promote health throughout the lifecycle.

2.3 Health Promotion

The approach on how one act in accordance to health and treatment of health can be separated in three distinguished phases. Treatment/rehabilitation stands for treating and relieving injury that is already emerged. Discovering means and interventions for removing risk factors for disease is the aim of preventional work. Whilst health promotional work is to bring awareness to behavior that maintain and encourage health by advancing skills that gives control over the environment and increase the energy level. In real-life health promotion and preventive work are interacting and

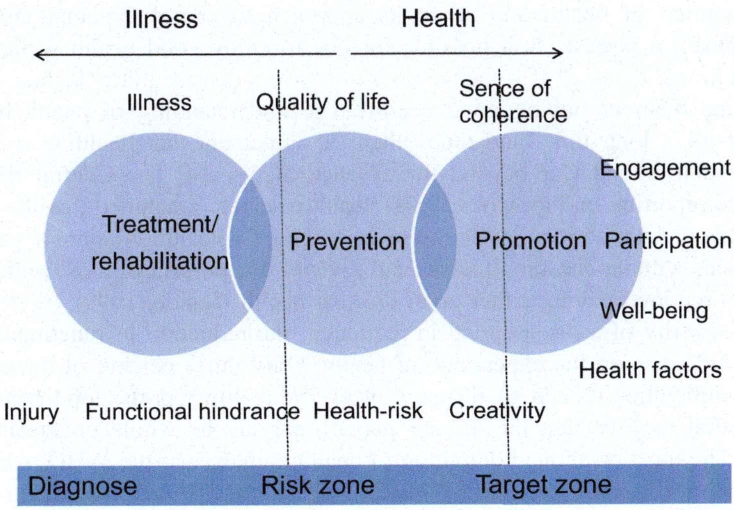

Fig. 2.2 Model of rehabilitation, prevention, and health promotion (Idébanken). *Source* Idébanken, http://www.idebanken.org/innsikt/artikler/helsefremmende-arbeidsplasser

complement each other, and different interventions can include the same elements (see Fig. 2.2).

WHO defines health promotion as "process of enabling people to increase control over, and to improve, their health" (WHO, 1986). The Ottawa charter for health promotion is an international agreement, elaborated and approved on the first international conference of health promotion, organized by WHO in Ottawa, Canada in 1986. Here they made an effort in supplementing WHO's definition by adding that health is a recourse in a humans' life, and not the object of the human existence. The Ottawa charter was the first document, which emphasized on health promotion and focused on making a conceptualization of health so that it could be more useable in health-related work and research. This was achieved when they realized which conditions had to be fulfilled for maintaining and achieving good health. In this agreement, it is stressed that health promotional work is closely associated with politics, all levels of the society and with the local environment. Through this, the aim was to achieve "health for everyone" within 2000. In Norway, one can, for example, see the effects of this through the Working Environmental Act, which in § 1-1 states that the act's purpose is to ensure a work environment that provides the basis for a healthy and meaningful working situation.

2.4 Definition of Health

As we can see from this discussion of its historical development, health has traditionally been defined as the absence of disease (Nelson & Simmons, 2003). In light of this understanding, one can ask questions about the definition and the understanding of health and about its approach to causal explanations for the development a disease. Is it possible for one to enjoy good health while dealing with a chronic disease? If someone does not have a disease but is feeling low, can we define them as having good health? The understanding of health has been debated for a long time, and today there is agreement that health is a complex concept that extends far beyond the absence of disease. The Central Bureau of Statistics reporting on Norway's (SSB) health research concluded that the absence of disease did not correlate with the main criteria for health. Ninety-six percent of individuals without chronic illnesses and seventy-one percent of those with chronic illnesses reported having either good or great health (Lunde, 1999).

The severity of a disease, and in particular, the reduction in functional ability, has an influence on the experience of health. Only thirty percent of those experiencing difficulties related to illnesses in everyday life reports good health. One explanation may be that the disease plays a part in the whole understanding of health. The most celebrated definition of health comes from the WHO, which sees health as a condition of entire physical, mental, and social well-being (WHO, 1948). This definition has been the subject of considerable criticism, especially because of the notion that absence of disease is a condition for obtaining good health and that good health is obtained through complete well-being. The challenges are based on three assumptions contained in the definition. Without exception, disease and health cannot coexist. Disease is disease; there is no room for differentiating between a mild and a more serious illness, such as cancer. The definition gives the impression that it is universal. The form it takes includes all humans without considering individual and social characteristics. Despite these criticisms, there is still value in this definition because it includes the notion of a positive well-being and gives the health concept a multidimensional character.

A definition of health that claims health to be more that the absence of disease gives rise to obstacles when defining and dealing with a positive health concept (Nelson & Simmons, 2003). Health has been described as a continuum between two end points where we find positive health on one side and disease on the other. It is not only medics that are interested in the definition of health today; the concept is also of great interest amongst philosophers. In their paper, Ryff and Singer (1998) compared different approaches from both a medical and a philosophical standpoint. The two disciplines have approached the concept of health in largely different ways. The focus of medicine has been disease, treatment, and the prevention of disease, whereas the focus of philosophy has been on well-being and mind–body interaction. Ryff and Singer's argument for the interaction between the mental and soma has been neglected for the last hundred years. They emphasize that having a goal in life, a social network, self-confidence, and the sensation of mastering your environment are important predictors for good health. Furthermore, the interaction between the mind and the body is important when we interpret and master obstacles within ourselves to obtain our physical health status. Ryff and Singer concluded that positive emotions contribute to a restitution and prevention of a disease. In this definition of positive health, they introduce three key principles: (1) Health is more of a philosophical question rather than a medical one, and therefore there is a need to articulate the good life, (2) well-being explains the relation between the mind and body in both mental and physical components and in how they affect each other, and (3) positive health is best understood as a multidimensional dynamic process rather than a scale that moves from bad health to good health. Interestingly, the discussion here might be framed by the question of whether health and disease are located at each extreme of the scale or whether they belong to two qualitative different scales. Positive psychology claims that qualitatively different results will be gained by analyzing positive health and negative health. Furthermore, they claim that by taking this research approach, new or additional predictors will be found for positive health. Below, we mention some of the studies that have examined this relationship.

Various empirical studies have focused on the concept of positive health. As early as 1994, Mackenbach, van den Bos, van de Mheen, and Stronks asked what were the health determinants for good and bad health, and whether these two conditions qualitatively differ from each other. Their study took an important step toward understanding health as a condition of well-being rather than being the lack of illness. Mackenbach, van den Bos, van de Mheen, and Stronks (1994) focused on socio-demographic variables that had earlier been used to predict disease, and they showed that these variables explained more of the variance in poor health than in good health. The researchers emphasized that the examined underlying variables were usually used in explaining the absence of disease and poor health. The results were explained within a model that was a better fit for examining poor health than it was for examining good health. The conclusion stated that the processes for generating good health have a lot in common with those that generate bad health; still, there is a distinct difference. Taking this into consideration, they recommended future research to examine other determinants, such as psychological, social, and cultural aspects, that consider the difference between pathogenesis and salutogenesis (Antonovsky, 1987).

In their study, Aronsson and Lindh (2004) argued that one could not understand factors that shape illness, discontent, and disorders. They suggested that by focusing on long-term healthy workers, one would identify areas for improving and intervening in the workplace, finding a common denominator to identify and improve constructive aspects. A study by Aronsson and Blom (2010) defined health using a behavioral-oriented or an action-oriented notion. They operationalised long-term health as a combination of low-level presenteeism and low levels of sickness absenteeism. They asked questions about which work-related and personal factors would increase the chances of workers remaining healthy. The results showed twenty-eight percent of the workers reporting being healthy throughout a long period of time. Furthermore, this study included other health benefit variables, such as management support, satisfaction with the work carried out, and clears goals. They also found that huge quantitative demands (e.g., workload and time pressure) hindering the satisfaction while working (Aronsson & Blom, 2010).

Research conducted on positive mental health suggests that mental health is a dynamic process that indicates more than the simple lack of disease. Seligman (2008) operationalised health as a combination of utmost status on each biological, subjective, and functional intention. Positive health is related to a "person–environment fit" kind of thinking, and an optimal adaptation between physical capacity, demands, and an individual's life resources. On its own, a positive mental health is seen as a goal in itself, as well as indicating a general health and high life expectancy. Seligman (2008) suggests that there is a logical explanation or rationale for studying positive mental health since it had earlier been a neglected field of study compared with the focus on poor mental health. The fundamental goal behind positive psychology lies in its close relation to health promotion, which is concerned with a change in the psychological focus toward building on strengths (Seligman, 2008). Seligman claims that when clinical illnesses come to an end, one does not simply end up with a positive mental health, including positive emotions, motivation, meaning, or positive relationships. Seligman (2008) mentions two main reasons for examining positive mental health in combination with poor mental health. First, humans continue to seek well-being while experiencing problems or/and disease. Second, the predictors of positive mental health, such as positive emotions, motivations, meaning, and positive relations, can protect against mental issues.

Another argument for shifting away from the "repairing and avoiding perspective" and toward a health promotion perspective is based on research that concludes that the medical objective of the search for disease does not show the expected significant association with the subjective health. The predictors used to explain poor health are failing when measuring subjective general health (Barsky, Cleary, & Klerman, 1992; Manderbacka, Lundberg, & Martikainen, 1999). Earlier research supports the notion of explaining more of the variance in subjective general health by using predictors related to both positive and negative health (Barsky et al., 1992). Several researchers agree on the need for more variables when researching the positive end of the health continuum (Aronsson & Blom, 2010; Aronsson & Lindh, 2004: Benyamini, Idler, Leventhal, & Leventhal, 2000; Mackenbach et al., 1994). Benyamini et al. (2000) examined other predictors for positive physical and psychological functioning, for example, engagement, positive affect, and social support. They found that the positive indicators were among the most important predictors for subjective health, both now and in the future. Happy and energetic people had a higher probability than others for maintaining a positive assessment of their health (Benyamini et al., 2000).

The discussion above has underlined the need to include both ends of the health continuum if we are to fully understand human subjective health. Equilibrium and balance are two qualitative different dimensions. Mind and body are not independent of each other, neither are they independent of the context around us.

2.5 Occupational Health

Health is closely related to function, and people tend to define their health in terms of how they function around their tasks. Here we find that occupational and health psychology deal with physical and mental health and well-being. The purpose of occupational health psychology is to develop, maintain, and promote workers health directly, as well as the health of their families (Quick et al., 1997). The earlier strong focus on negative factors concerning work (demands, stress, and lack of motivation) has more recently been balanced with a greater attention placed on the positive factors concerning occupational health, such as a focus on resources, well-being, and motivation (Bauer & Jenny, 2012). Today, we find a growing agreement on the need to include both demands and resources in understanding the underlying processes.

2.6 Conclusion

In conclusion, it is necessary to reflect on the important question of who is in charge of our health. Is it the individual, or is it the society? This is a difficult, but an important political discussion. We know that lifestyle has an important impact on health, and hence we have some control over improving our health, such as through diet and exercise. We also know that health is influenced by social differences, something that is explained by determinants from social positions. The strong focus on lifestyle, as well as the individual's behavior, has been criticized as a search for scapegoats. People are blamed for behavior over which they have little or no control (Fugelli & Ingstad, 2009). The line between individual and society is unclear, and it is less likely that individual changes can rise without the society also changing. This view was agreed upon and became dominant in the World Health Organisation's first world conference with health promotion on the agenda. The conclusion noted a reciprocal commitment. The aim of this chapter has been to capture the complexity in the notion of health; when faced with difficulties in society, the importance of understanding this complexity is important, both for the employers and, not least, the employees.

References

Antonovsky, A. (1987). *Unravelling the mystery of health: How people manage stress and stay well*. San Fransisco: Jossey-Bass.

Aronsson, G., & Blom, V. (2010). Work conditions for workers with good long term health. *International Journal of Workplace Health Management, 3*(2), 160–172.

Aronsson, G., & Lindh, T. (2004). *Långtidsfriskas arbetsvilkor - En populationsstudie* [Long term health at work—a study of the population]. Arbete och Hälsa: 10. Stockholm: Arbetslivsinstitutet.

Barsky, A. J., Cleary, P. D., & Klerman, G. L. (1992). Determinants of perceived health status of medical outpatients. *Social Science and Medicine, 34*(10), 1147–1154.

Bauer, G. F., & Jenny, G. J. (2012). *Salutogenic organisations and change*. Heidelberg: Springer.

Benyamini, Y., Idler, E. L., Leventhal, H., & Leventhal, E. A. (2000). Positive affect and function as influences on self-assessments of health: Expanding our view beyond illness and disability. *Journals of Gerontology. Series B, Psychological Sciences and Social Sciences, 55*(2), 107–116.

Engel, G. L. (1977). The need for a new medical model: A challenge for biomedicine. *Science, 196* (4286), 129–136.
Espnes, G. A., & Smedslund, G. (2009). *Helsepsykologi [Health psychology]*. Oslo: Gyldendal Norsk Forlag.
Fugelli, P., & Ingstad, B. (2009). *Helse på norsk – god helse slik folk ser det* [Health in Norwegian—Good health the way people see it]. Oslo: Gyldendal akademisk.
Hem, K.-G. (2011). *Bedriftenens kostnader ved sykefravær. En undersøkelse blant NHO- bedrifter* [Costs of an organization when sickness absence—A survey among NHO-organisations]. SINTEF-rapport A19052. Oslo: SINTEF Teknologi og Samfunn.
Lunde, E. S. (1999). *Helsetilstanden i befolkningen* [The state of health in the population]. Notater 33–1999. Oslo: Statistisk Sentralbyrå.
Mackenbach, J. P., Van Den Bos, J., Van De Mheen, H., & Stronks, K. (1994). The determinants of excellent health: Different from the determinants of ill-health? *International Journal of Epidemiology, 23*(6), 1273–1281.
Manderbacka, K., Lundberg, O., & Martikainen, P. (1999). Do risk factors and health behaviours contribute to self-ratings of health? *Social Science and Medicine, 48*(12), 1713–1720.
Matrazzo, J. D. (1980). Behavioral health and behavioural medicine: Frontiers for a new health psychology. *American Psychologist, 35*(9), 807–817.
Nelson, D. L., & Simmons, B. L. (2003). Health psychology and work stress: A more positive approach. In J. C. Quick & L. E. Tetrick (Eds.), *Handbook of occupational health psychology* (pp. 97–119). Washington, DC: American Psychological Association.
Quick, J. C., Camara, W. J., Hurrell, J. J., Johnson, J. V., Piotrkowski, C. S., Sauter, S. L., et al. (1997). Introduction and historical overview. *Journal of Occupational Health Psychology, 2* (1), 3–6.
Ryff, C. D., & Singer, B. (1998). The contours of positive human health. *Psychological Inquiry, 9* (1), 1–28.
Seligman, M. E. P. (2008). Positive health. *Applied psychology: An International Review, 57,* 3–18.
Suls, J., & Rothman, A. J. (2004). Evolution of the psychosocial model: Implications for the future of health psychology. *Health Psychology, 23*(2), 119–125.
World Health Organisation. (1948). *World Health Organization constitution*. Geneva: WHO.
WHO Ottawa. (1986). Report of an International Conference on Health Promotion, 1986, Ottawa, Ontario, Canada 17–21 November 1986. *Health Promotion: An International Journal*, 1:4 i–v, 405–460.

Chapter 3
Explanatory Models in Occupational Health Psychology

Per Øystein Saksvik

There are a number of models that are designed to provide insight into the correlation between the concepts that are usually included in our understanding of the relationship between occupation and health. The most famous models have been used in numerous empirical studies and have received diverse support. Among the concepts that recur in several of the models, "demands of the job" has been solidly established as an important factor, even though it has been discussed whether a quantitative claim (e.g., high tempo) or a qualitative claim (e.g., difficult customers) is most influential concerning health. Which factors are most influential also has a relation to how the workforce develops from an industry-dominated to knowledge-dominated working life.

This chapter seeks to systematize different models that are in play when explaining occupational health psychology. The idea is not to give a thorough introduction to each and every model; rather, the intention is to find which principles are embedded in the different models. Based on hallmarks of the different models, they can be sorted according to three different perspectives: balance, harmony, and conflict. A weighing scale that seeks equilibrium and thus being correctly aligned, illustrates the perspective of balance. The perspective of harmony is based upon relationships that build upon one another and become an ideal state, such as when building blocks make a framework. The conflict perspective illustrates forces that oppose each other, and ultimately affect a desirable development, not by building on each other but by competing until the best solution is provided. The description of these models will give a further understanding of what is meant by the different perspectives.

A representative of the perspective of balance is Karasek and Theorell's demands-control-support (DCS) model with its historical roots dating back to 1979

P.Ø. Saksvik (✉)
Department of Psychology, Norwegian University of Science and Technology (NTNU), Trondheim, Norway
e-mail: per.saksvik@ntnu.no

M. Christensen et al. (eds.), *The Positive Side of Occupational Health Psychology*,
https://doi.org/10.1007/978-3-319-66781-2_3

(Karasek, 1979; Karasek & Theorell, 1990). It was inspired by Robert Karasek's time spent in Sweden, where he encountered Gardell's works (Sørensen, Hasle, Hesselholt, & Herbøl, 2012). Gardell helped provide an understanding of the concepts of influence, participation, and democracy in the workforce (Sørensen et al., p. 45). Until this day, the DCS model has most frequently been used when researching working environments and health. It is probably still the most frequently used model but has received competition from the job demand–recourses (JD-R) model (Demerouti, Bakker, Nachreiner, & Schaufeli, 2001). This model has elements from Karasek's model embedded in it, especially the demands. After presenting the DCS model, we shall return to an explanation of the JD-R model.

Karasek's (1979) original model defined the psychosocial working environment based on two dimensions—psychosocial job demands and control. The quality of the psychosocial working environment is decided based on the how workers' levels of demands and control are relative to each other. According to Karasek and Theorell (1990), high job demands will have a negative effect on workers only in those situations where the workers feel a lack of control and a lack of participation in decisions concerning them. A sense of control gives workers greater opportunities to use their set of skills and to contribute to developing an expertise, thus giving the feeling of mastering the job. This again develops individual skills and makes it easier to overcome high demands and challenges. In other words, one finds oneself in an optimal working situation when high demands are combined with a sense of control, that is when the two parameters are in balance. This combination will lead to psychological growth and learning and an extension of one's skill.

It is well known that this balance only has an optimal effect when the demands and the control are high. When the demands are few and the control low, one gets passive activity. Passive activity causes a loss of skills and reduces the ability to master challenges. The most negative imbalance in Karasek and Theorell's model is called iso-strain. This situation is characterized by high demands combined with low job control and low social support, and therefore, the worker finds it hard to maneuver the demands. The result is psychological stress; this can again result in health problems and contribute to sickness absenteeism.

3.1 Job Control and Collectivism

The concept of control has had long-term effects on Norwegian working life research. One of the first people to work with this notion in working life was Bertil Gardell in Sweden (Sørensen et al., 2012). Later, the concept had success in the world, particularly thanks to Karasek who was inspired by the Swedish research. The control dimension (decision authority) is based upon a combination of two theoretical separate conceptions which Karasek and Theorell (1990) claim have a close practical connection: firstly, the degree to which the workers are exploited and develop different skills or knowledge (capacity utilization) and secondly, the authority to make decisions about their own job. The mutual strong relationship between these subcomponents is based on the acquisition of skills that give the workers influence over the job process. Karasek and Theorell emphasize giving workers a high knowledge or skill level that provides workers with control over which specific skills they need to acquire for a task.

The impact of this concept had through Karasek's model has been important, but working life has changed largely since the 1970s and 1980s, and it appears that the concept is about to lose its position (Eiken & Saksvik, 2011). The transition from an industrial society to a knowledge-based society has played an important role. Karasek (and Theorell) developed the job content questionnaire (1985) and with it a notion of individualistic orientation, although its implicit claim is that control is related to something beyond the individual. One example of a question from Karasek is as follows: "How often do you have any influence on what you do at work?" The way a job is organized in Scandinavian countries has not only been the concern of the single worker. Management and unions, in cooperation, have defined jobs. The workers have to a great degree delegated their opinion to their representative from the union who negotiates with the management on behalf of the collective. In this way, there is duplicity in this sort of question. Someone would report having this kind of influence through the negotiation, while others, on a larger scale, would answer with their own point of view. Still, others would think the question irrelevant either because external owners (foreign companies) define the job or because they are self-employed (or largely independent) and make their own decisions without concern for the "influence on what you do at work." The complexity in today's working life makes it difficult to define what control over the job means, but that something collective in the working life exists, something beyond the single individual, is still at play. But what is it?

The workers' collective has a sound history in Norwegian and Nordic working life. The sociologist Lysgaard (1967) brought the idea into play in his studies at M. Peterson and Søn A/S Cellulose in Moss, Norway. Lysgaard wished to research the relationship between workers and management from below, that is, from the workers' perspective. He was concerned with the paradoxical relationship in which the workers share a common fate with the organization of which they are a part. As long as the organization survives, the workers survive as well. But they are also in a state of discord with the organization because they are subject to the organization's technical and economic system and become "tools of the production" in this system without having any voice concerning the system (Sørensen et al., 2012).

The workers' collective is a social system that arises spontaneously and seeks to take care of humans in the technical/economic system, which appears insatiable, one-sided, and ruthless (Sørensen et al., 2012). The workers' collective becomes a social mechanism for accommodating the single worker's personal interests in figuring out the power balance, economic interests, workload, and social status. This kind of social mechanism becomes informal in contrast to the formal technical/economic system and constitutes of a set of norms concerning social relations.

Lysgaard's main contribution was never translated into English. Therefore, it is interesting that in the 1980s a term called "collective control" appeared in American research (Johnson, 1989). We do not know if there has been any contact between

Norwegian and American researchers (besides Karasek), which could have inspired Johnson or if it could be a parallel development. Johnson emphasized the social dimension of control and used the term collective control in his research on the health effects of the workers' control over the job process, where the combination of the workers' control and the job process and the social relations on the job became a strategy for the workforce's survival (Saksvik, Hammer, & Nytrø, 2013). Collective control here becomes a concept in the formal relational system between each part concerning power, negotiation, and distribution. This, in turn, becomes the main difference between Johnson, and Lysgaard, as Lysgaard emphasized the informal social system.

Emphasis on the informal part of the social system through the social norms of the job (Saksvik et al., 2013) is a position we have also acquired. Social norms are informal but still concrete enough to be measured. We can form our own norm scale consistent with conceptions shared among workers in the job (Hammer, Saksvik, Nytrø, Torvatn, & Bayazit, 2004). The shared conceptions implicate that the question should be whether more people than you share the same thought. This indicates that there is a common conception among the workers and that every single one is informed of this fact. Here, we get a new question formulated such as, "In the job is our communication so good that most of the misconceptions are sorted out?" By drawing the attention toward norms or a common conception, we present our norm scale with this heading: "Now follows questions or statements about some common conceptions among the workers in your company." It is not immediately assumed that all answers are solemnly answered based on the community's interests and attitudes regarding these norm questions. Some would rather emphasize their own attitudes toward a phenomenon and create a mixture. With this in mind, it could be easier to ask only an individual and then compute how much of the variance in one question about the individual is explained by using methods such as hierarchy linear modeling (HLM). This is a method that calculates both the individual variance and the community's variance (the group, unit, or collective) (Saksvik et al., 2013). When you then know which unit, department, and/or company to whom the individual belongs, one can calculate how much of the variance is explained by common conceptions. In our studies, we have seen that between 4 and 20% of the variance (depending on which variable we study) is explained by the individual level (Hammer et al., 2004; Rennesund & Saksvik, 2010).

3.2 Social Control

Collective control consists of a common experience in the workers' environment. An adjoining concept that also has received a lot of attention in research is social support. Some will see this as overlapping, especially if we talk about collegial support. From a historical view, House introduced the concept in 1981 (House, 1981), and there is no doubt this is a meaningful factor between stress and strain and social support and the impact, and it has on cardiovascular diseases and sickness absenteeism (Karasek & Theorell, 1990; Shumaker & Czajkowski, 1994). Karasek and Theorell (1990) have noted as well the negative consequences of working in social isolation, where workers are hindered from contacting colleagues (called iso-strain). If we take a closer look at the operationalization that Karasek used (1985), there is a separation between management support and collegial support. The management's support has a paternalistic angle, with the questions expressing a top-down attitude (help and support). Collegial support is operationalized as well in the job content questionnaire as help support, i.e., as a sort of friendly favor. House (1981) differentiated the term "social support" into four dimensions: (1) emotional support that gathers empathy, care, love, and trust; (2) instrumental support, related to materials and practical assistance; (3) informational support which concerns advice that can be useful when you have problems; and (4) assessment support that includes constructive feedback and confirmation of oneself. Karasek's (and Theorell's) expansion of the model to include social support as a third dimension is based on the contribution of House (Karasek & Theorell, 1990).

Another adjoining concept that is somewhat connected with the previous social support is social capital. Hasle and Møller (2007) showed how a culture based on conflict and different interests was less expedient than a culture based on common support and respect, as a result of the acquired social capital. Oksanen, Kouvonen, Kivimäki, and Pentti (2008) showed that an increasing social capital was related to lasting good health. Social capital as a concept has grown to take a part in the cultural perspective of an organization. Putnam has defined social capital as the sum of norms, support, and network (Putnam, 1988), which is a quite comprehensive definition. Putnam's definition sets the level of support and the experience of justice in an organization as an expression of the business's ability to cooperate. This is based on three relational assessments: bonding (within a unit or department), bridging (between a unit or department), and linking (between the management and workers).

It is, therefore, not an easy division between the control-based and the social dimension among the relationships on the job. We have claimed that the collegial control expresses a somewhat more basic relational dimension, a sort of solidarity, which develops as a socially constructed phenomenon in a social community through social interactions and interpersonal relations (Saksvik et al., 2013). Collective control now becomes something more than and different from social support but still remains a part of the social dimension of the job. Research indicates that when the management feels social support, they react by giving more support back (Shanock & Eisenberger, 2006). This way, a supportive organizational culture can be created through a positive spiral where both the management and workers perceive support and give it to each other. If the support is embedded as a top-down measure, it can feel like imposed social support and be viewed as a negative and stress-inducing effect on those who are receiving it (Deelstra et al., 2003).

3.3 The Model of Effort–Reward Imbalance

Siegrist (1996) purposed the model of effort–reward imbalance (ERI model) when explaining work-related stress and gave an alternative to the demand–control–support model. In this model, the basic thought is that the individual's expectation of a social life in the community can be reflected in the job, and thus an imbalance between effort and reward becomes straining and results in stressful actions. The effort is a result of external demands (such as time) or internal demands (the individual motivation, ambitions, etc.), and Siegrist (2002) identified that there is a conceptual overlap between these variables and demands, which we see in the demand–control–support model. The focus shifts from control and toward the variable reward, which can be present in the form of money, recognition, or status control (Siegrist, 1996).

The ERI model highlights that balance is in relation to health and well-being. High effort and low reward is seen as the most harmful combination (Siegrist, 2002). He also includes overcommitment as a factor that triggers imbalance since it involves the workers being subject to demands with a higher motivation and expectation of oneself, expecting more responsibility and working harder to keep deadlines. Siegrist expected this factor to give an extra dimension in relation to the demand–control–support model, which he meant only included factors related to the job and did not include personal factors. It is thought that the most evident negative health effect will present itself when the worker is faced with an imbalance combined with overcommitment. On the other note, Siegrist (1996) highlights that imbalance can be maintained for a while without it being harmful, but only when the worker sees the beneficial long-term effects.

Cardiovascular health was an early focus when it came to consequences of the effort–reward imbalance, and there was evidence pointing toward a common condition (Siegrist, 1996). In a later literature analysis of 45 empirical studies, van Vegchel, de Jonge, Bosma, and Schaufeli (2005) noted that effort–reward imbalance had a significant association with others health outcomes, such as psychosomatic health and job-related well-being. This study concluded that the model gives a good understanding when it comes to relations between the job and health, and it could, therefore, be a better-suited predictor for general health effects than a more specific condition.

3.4 The Job Demands–Resources Model

Another balance model is job demands–resources (JD-R), which as with JDCS and ERI aims to explain stress as a consequence of an imbalance between two factors. One factor is job demands, which are physical, psychological, social, or organizational aspects of the job that demand maintenance of one's physical and/or psychological effort or ability, and thus are connected with physiological or psychological costs for the individual (Bakker & Demerouti, 2007). On the other hand, are job resources which represent physical, psychological, social, or organizational aspects of the job which contribute to goal achievement, reduction in demands and the physiological and psychological demands associated with them, and/or stimulate personal growth, learning, and development (Bakker & Demerouti, 2007). Schaufeli and Taris (2014) discussed how JD-R, combined with JDCS and ERI, are more usable in job-related settings because of the general definitions of the essential variables job demands and job resources. The theory suggests a negative correlation between job demands and job resources, and furthermore, job demands contribute to strain for the individual, while job resources contribute to motivation. In addition, job resources work as a buffer against strain as a result of job demands. The strain due to high job demands will further contribute to adverse outcomes such as stress and burnout, while motivation contributes to positive outcomes (see Chap. 7 for more information about the JD-R-model) (Fig. 3.1).

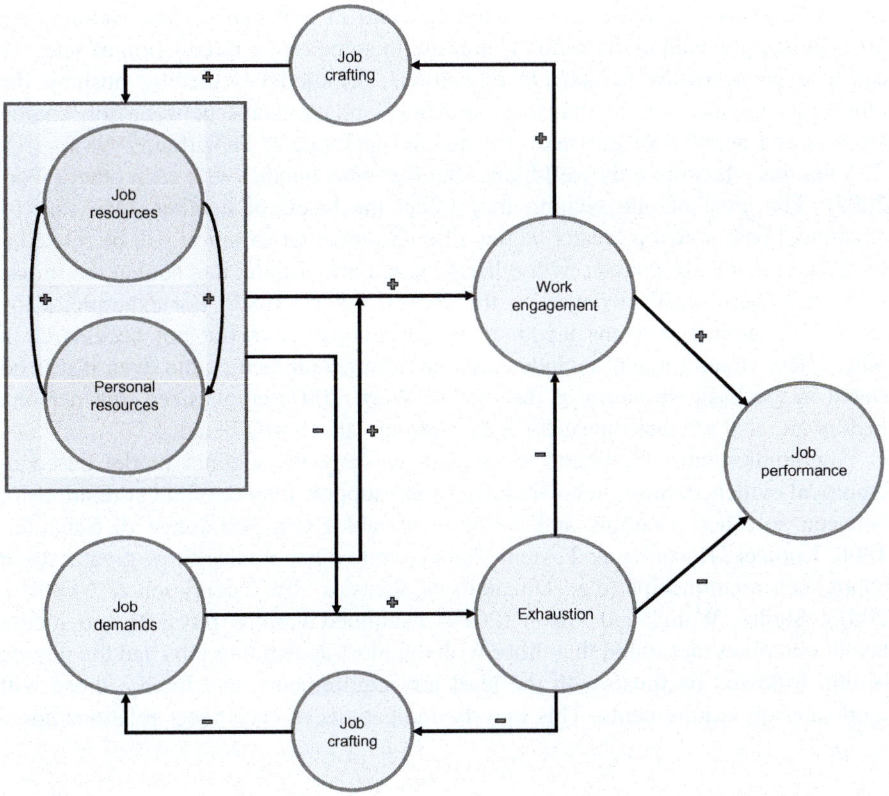

Fig. 3.1 Job demands–resources model (JD-R). *Source* Bakker and Demerouti (2014), © Wiley-Blackwell

3.5 The Harmony Perspective

Warr (1987) proposed a framework to explain well-being in a work context as well as in other social structures. The model differs in that the original was supposed to explain well-being, a positive outcome, unlike many other theories which originally focused mostly on explaining stress and negative health effects. Initially, the background for the theory was that nine factors have an impact on health. These factors are the opportunity for personal control, the possibility of knowledge utilization, externally generated goals, variation, environmental predictability, contact with others, the availability of money, physical security, and appreciation of social position. Later, he also included the three factors: supportive supervision, good career prospects, and fair treatment, which he believed was essential in a work context (Warr, 2007). Warr said that the relationship between the factors and wellness is analogous to the effect vitamins have on physical health, and the model was, therefore, called the vitamin model. The assumption is that the well-being increases with the increased presence of the factors, or vitamins. When reaching a certain level, however, the effect stagnates, and in terms of certain vitamins can growing supply even be harmful. Therefore, in some cases, a reduction of vitamins can also be necessary to achieve relaxation. The model is thereby pushing the, otherwise, popular notion that there are linear relationships between job characteristics and health variables like well-being (de Jonge & Schaufeli, 1998).

Vitamins affect not only well-being, but they also interact with each other (Warr, 2007). The level of one vitamin may affect the levels of another. One can, for example, imagine that a greater degree of environmental variation will be related to the greater degree of contact with other people. Furthermore, it is so that the impact of a vitamin on health depends on the context (Warr, 1987). For example, a low level of a vitamin worsens health in some circumstances but not necessarily in others. The vitamin model includes only environmental factors, but even if it is not stated in the basic structure of the model, Warr (2007) emphasized that personal factors are also of importance for how vitamins affect well-being.

Few studies have facilitated to confirm or deny the vitamin model based on empirical evidence. Some scholars have found support for a curvilinear relationship between so-called vitamins and wellness variables (e.g., de Jonge & Schaufeli, 1998; Kubicek, Korunka, & Tement, 2014), while other studies show greater doubt about such relationships (e.g., Donatelli & Sevastos, 2007; Jeurissen & Nyklicek, 2001). Shultz, Wang, and Olson (2010) examined various physical and mental health outcomes and found that those with the most demanding jobs had the poorest health, followed by those with the least job requirements, and finally, those with moderate job requirements. This may be indications of curvilinear relationships.

3.6 The COR Theory

A different theory which can be categorized in what we call the harmony perspective, is Hobfoll's (1988, 1989) motivational theory, the model of conservation of resources (COR). This model was proposed as an alternative to models like JDCS, and the aim was to explain the process around work-related stress based on one single factor resources. The basis of the model is the assumption that all humans have a driving force to create, foster, preserve, and protect the quantity and quality of their recourses. Recourses are defined as objects, personal characteristics, circumstances, or energies that are valued by the individual (primary resources), or which may function to obtain additional resources (secondary resources).

Based on the COR model, stress can occur when one is faced with a loss of resources or one actual experiences loss of resources (Hobfoll, 1989). Also, the individual has to invest resources when protecting the resources one possesses and obtain new ones. A third source of stress arises, therefore, if one does not obtain the expected gain from the investments; this can be interpreted as if also the theory contains an aspect of the balanced perspective. Hobfoll (2001); however, points out that it is the changes, and not in the stability in the resource levels, that influence people's health is the essence of the COR theory.

The value of the COR theory was strengthened when the interest in job engagement and positive psychology increased and became, in the aftermath of the JD-R proposal, a theoretical explanation in the model for some of the correlations (Gorgievski & Hobfoll, 2008). Resources have not only had an impact on stress and on the further consequences of burnout but also on job engagement and well-being. For example, loss of resources will hinder the process that leads to job engagement, while the presence and gain of resources make it possible to maintain and obtain job engagement.

Besides the already mentioned, so-called loss and gain spiral central to the COR theory (Gorgievski & Hobfoll, 2008), a loss spiral can be a consequence of one losing resources and not having the sufficient resources to avoid additional losses. In other words, one has an increased risk of losing more resources, the fewer one already has. On the other side, one has a gaining spiral, whereby the possession of resources makes it easier to obtain new ones. This is because an individual is more likely to have the resources needed to avoid losses and can also "afford" to invest resources in achieving further gain. Additionally, the possession of great resources has a tendency to bring about more resources, which can develop to become something called resource caravans. These tend to build up during one's life and are relatively stable. This suggests that resources caravans also will promote a more stable condition concerning health and well-being since they make resource losses less visible.

3.7 The Conflict Perspective

The conflict perspective has, as far as it is known, not been used in occupational health psychology. Therefore, we are moving toward the organizational change literature and using a model developed by Burns (2004). The model was inspired by chaos theory and concerns how an organization should grow and develop. The basic idea of the model makes it the complete opposite of the balance and harmony perspective. This idea claims that stability and calmness (labeled ossification) lead to stagnation and, in the worst cases, death. Imbalance and disharmony (labeled dynamic turbulence) lead, on the other hand, to basic values being challenged and the organization being driven forward (Saksvik & Tvedt, 2009). It may be that the name of such a mindset should rather be constructive conflicts (Tvedt, Saksvik, & Nytrø, 2009). The conflict has a negative ring in practical relations, but the thought behind the phenomenon is that it contributes to positive development through the good forces winning and securing progress. When a simple physical phenomenon's drip from a faucet cannot be explained mathematically, what about the complex human systems then? It is not a discussion of randomness in a system but rather complexity. Chaos is paradoxical because short-term patterns are hard to reveal, while long-term patterns in an organization are easily discovered in retrospect. An organization always moves between stability and movement in relation to a dynamic surrounding [i.e., must be capable of both single-loop and double-loop learning at the same time (Argyris & Schön, 1996)]. There must never be the fear of failure in an organization since it prevents creativity and development.

3.8 Measuring Occupational Health Psychology

Evaluation of occupational health psychology models happens usually by means of questionnaires given to the employees to uncover their attitudes regarding different areas. These are attitudes that can be related to a single personality (dispositions), available resources in the work, relations with colleagues or leadership, or abilities to master and change. Constructive conflict perspective can also be evaluated on an individual level through many of the same questions, perhaps, in particular, the questions concerning openness, ability to learn from mistakes, the conception of the leaders, and the ability to change. Meanwhile, there is an important difference between traditional models and constructive conflict models; that is the time dimension. Development happens over time, and an evaluation has to consider this fact. It is, thus, not enough to survey the available resources, abilities, and attitudes at a particular time, but it is necessary to follow the same individual, working group, and organization over a longer period of time. Nor is there something in the conflict theory that suggests that development happens smoothly and continuously. Changes may rapidly force their way through as a result of an external impact that occurs periodically. Basic capacities, however, may indicate that the individual, working group, and organization are on the right course. These capacities may be measured at any time, and as mentioned, it will involve measuring the ability to learn from past mistakes, the willingness to change, availability of the necessary resources for development, and the evaluation of the leader's competence to lead in the right direction. Several of these measurements do already exist but are not compiled to measure progress and to change potentiality on an individual or an organizational level. Additionally, an even greater precondition for obtaining a good measurement in constructive conflict perspective rather than in the traditional models is to conduct longitudinal studies, preferably over longer periods of time than those which are conducted now, i.e., of a double-digit level. Of course, whether this is realistic or not should be discussed. The use of such a scale can show great fluctuations over time, depending upon whether the organization is in an insecure phase in which there are great refractions and many challenges, or if it is in a more constructive phase, in which everything works. This is common and to obtain a constructive developmental progress, the organization has to face some lows so that the organization and the individuals in it can reorient themselves find new solutions and implement new initiatives. A problem arises if they never recuperate from the low phase, then there is an imminent danger that the organization in its current form has lost its right to exist. As we know, this also happens regularly. There are not many organizations that remain unchanged over decades.

We have developed a few scales that are included in the measuring of constructive conflict. The scale to measure constructive conflict by Health Change Process Index (HCPI) (Tvedt et al., 2009) is based on questions about the willingness to change. It measures resistance to change collected from Oreg (2003), communication in the organization (Downs & Hazen, 1977), and optimism (Scheier, Carver, & Bridges, 1994). We have also included demands of the job (Karasek, 1985, for example: "Does your job demand that you obtain new knowledge and skills?" This is a question included in Karasek's JCQ, but not in the questions most commonly used. In order to develop a measuring device that considers design and growth seriously, it is important that these issues about resources are raised (Pejtersen, Kristensen, Borg, and Bjørner, 2010), for example: "The conflicts are solved in a righteous manner," and that they measure the innovative climate as in the healthy organization barometer (Lindström, Hottinen, & Bredenberg, 2000).

3.9 The Model in the Nordic Perspective

A model of constructive conflict has had an evident association with the Nordic perspective of work life. Through the decades, collaboration has been the key factor for the development of working life in Scandinavia, and a key factor here is that the parties challenge each other and discuss strategies for the future (Sørensen et al., 2012). This takes place both at a political level and in every company or business. The foundation according to the model of Burns (2004) is precisely that the leadership and the employees engage in a dialog regarding the worldview and possibilities for the future. The perspectives have to be challenged, and if this does not happen, the organization may stagnate and risk getting into a difficult situation on an occupation health psychology level which can lead to negative consequences for the employees in the form of stress, burnout, sickness absenteeism, and turnover intentions that can both affect the organization and the employee negatively. This is a typical intersection between organizational and occupational psychology. Increased organizational prevention through thinking long term is also a positive investment in the employees' health.

References

Argyris, C., & Schön, D. A. (1996). *Organizational learning II*. USA: Addison-Wesley Publishing Company.

Bakker, A. B., & Demerouti, E. (2007). The job demands-resources model: State of the art. *Journal of Managerial Psychology, 22*(3), 309–328. doi:10.1108/02683940710733115.

Bakker, A. B., & Demerouti, E. (2014). Job demands-resources theory. In P. Y. Chen & C. L. Cooper (Eds.), *Work and wellbeing: Wellbeing: A complete reference guide* (Vol. III, pp. 37–64). Chichester, UK: Wiley-Blackwell.

Burns, J. S. (2004). Chaos theory and leadership studies: Exploring uncharted seas. *Journal of Leadership & Organizational Studies, 9,* 42–56.

de Jonge, J., & Schaufeli, W. B. (1998). Job characteristics and employee well-being: A test of Warr's vitamin model in health care workers using structural equation modelling. *Journal of Organizational Behavior, 19,* 387–407.

Deelstra, J. P., Peters, M. C., Schaufeli, W. B., Stroebe, W., Zijlstra, F. R., & van Doornen, L. P. (2003). Receiving instrumental support at work: When help is not welcome. *Journal of Applied Psychology, 88*(2), 324.

Demerouti, E., Bakker, A. B., Nachreiner, F., & Schaufeli, W. B. (2001). The job demands-resources model of burnout. *Journal of Applied Psychology, 86,* 499–512.

Donatelli, F., & Sevastos, P. (2007). Wellbeing at work: A multivariate analysis of Warr's vitamin model. In A. I. Glendon, B. M. Thompson, & B. Myors (Eds.), *Advances in organizational psychology*. Brisbane: Australian Academic Press.

Downs, C. W., & Hazen, M. D. (1977). A factor analytic study of communication satisfaction. *Journal of Business Communication, 14*(3), 63–73. doi:10.1177/002194367701400306.

Eiken T. E., & Saksvik P. Ø. (2011). Det psykososiale arbeidsmiljøet—hvor har det blitt av den sosiale dimensjonen? [The psychosocial work environment—what has become of the social dimension?] In P. Ø. Saksvik (Ed.), *Work and organizational psychology—Current topics of inspiration for a better working life*. Oslo: Cappelen Akademisk Forlag.

Gorgievski, M. J., & Hobfoll, S. E. (2008). Work can burn us out or fire us up: Conservation of resources in burnout and engagement. In J. R. B. Halbesleben (Ed.), *Handbook of stress and burnout in health care*. Hauppauge, NY: Nova Science.

Hammer, T. H., Saksvik, P. Ø., Nytrø, K., Torvatn, H., & Bayazit, M. (2004). Expanding the psychosocial work environment: Workplace norms and work-family conflict as correlates of stress and health. *Journal of Occupational Health Psychology, 9,* 83–97.

Hasle, P., & Møller, N. (2007). From conflict to shared development: Social capital in a tayloristic environment. *Economic and Industrial Democracy, 3,* 401–429.

Hobfoll, S. E. (1988). *The ecology of stress*. Washington, DC: Hemisphere.

Hobfoll, S. E. (1989). Conservation of resources: A new attempt at conceptualizing stress. *American Psychologist, 44*(3), 513–524.

Hobfoll, S. E. (2001). The influence of culture, community, and the nested-self in the stress process: Advancing conservation of resources theory. *Applied Psychology, 50*(3), 337–421. doi:10.1111/1464-0597.00062.

House, J. S. (1981). *Work, stress and social support*. Reading, MA: Addison Wesley.

Jeurissen, T., & Nyklicek, I. (2001). Testing the vitamin model of job stress in Dutch health care workers. *Work & Stress, 15*(3), 254–264. doi:10.1080/02678370110066607.

Johnson, J. V. (1989). Collective control: Strategies for survival in the workplace. *International Journal of Health Services, 19,* 469–480.

Karasek, R. A. (1979). Job demands, job decision latitude, and mental strain: Implications for job redesign. *Administrative Science Quarterly, 24,* 285–308.

Karasek, R. A. (1985). *Job content questionnaire, Revision 1.1*.

Karasek, R. A., & Theorell, T. (1990). *Healthy work: Stress, productivity and the reconstruction of working life*. New York: Basic Books.

Kubicek, B., Korunka, C., & Tement, S. (2014). Too much job control? Two studies on curvilinear relations between job control and eldercare workers' well-being. *International Journal of Nursing Studies*. doi:10.1016/j.ijnurstu.2014.05.005.

Lindström, K. V., Hottinen, V., & Bredenberg, K. (2000). *The healthy organization barometer* [*Tyoilmapiiri-ja hyvinvointibarometri*]. Helsinki, Finland: Työterveyslaitos.

Lysgaard, S. (1967). *Arbeiderkollektivet* [*Workers' collective*]. Oslo: Universitetsforlaget.

Oksanen, T., Kouvonen, A., Kivimäki, M., & Pentti, J. (2008). Social capital at work as a predictor of employee health: Multilevel evidence from work units in Finland. *Social Science and Medicine, 3,* 637–649.

Oreg, S. (2003). Resistance to change: Developing an individual difference measure. *Journal of Applied Psychology, 88,* 680–706.

Pejtersen, J. H., Kristensen, T. S., Borg, V., & Bjørner, J. B. (2010). The second version of Copenhagen Psychosocial Questionnaire (COPSOQII). *Scandinavian Journal of Public Health, 38*(suppl. 3), 8–24.

Putnam, R. D. (1988). Diplomacy and domestic politics: The logic of two-level games. *International Organization, 42*(03), 427–460.

Rennesund, Å. B., & Saksvik, P. Ø. (2010). Work performance norms and organizational efficacy as cross-level effects on the relationship between individual perceptions of self-efficacy, overcommitment, and work-related stress. *European Journal of Work and Organizational Psychology, 1,* 629–653.

Saksvik, P. Ø., Hammer, T. H., & Nytrø, K. (2013). Social relations at the collective level: The meaning and measurement of collective control in research on the psychosocial work environment. *Nordic Journal of Working Life Studies, 3*(3), 30–46.

Saksvik, P. Ø., & Tvedt, S. D. (2009). Leading change in a healthy way. *Scandinavian Journal of Organizational Psychology, 1,* 20–29.

Schaufeli, W. B., & Taris, T. W. (2014). A critical review of the Job Demands-Resources model: Implications for improving work and health. In: G. Bauer & O. Hammig (Eds.), *Bridging occupational, organizational and public health* (pp. 43–68).

Scheier, M. F., Carver, C. S., & Bridges, M. W. (1994). Distinguishing optimism from neuroticism (and trait anxiety, self-mastery, and self-esteem): A reevalutation of the Life Orientation Test. *Journal of Personality and Social Psychology, 67*(6), 1063–1078.

Shanock, L. R., & Eisenberger, R. (2006). When supervisors feel supported: Relationships with subordinates' perceived supervisor support, perceived organizational support, and performance. *Journal of Applied Psychology, 9*(3), 689.

Shultz, K. S., Wang, M., & Olson, D. A. (2010). Role overload and underload in relation to occupational stress and health. *Stress & Health, 26,* 99–111.

Shumaker, A. S., & Czajkowski, S. M. (1994). *Social support and cardiovascular disease*. New York, NY: Plenum Press.

Siegrist, J. (1996). Adverse health effects of high-effort/low-reward conditions. *Journal of Occupational Health Psychology, 1*(1), 27–41.

Siegrist, J. (2002). Effort-reward imbalance at work and health. In P. L. Perrewé & D. C. Ganster (Eds.), *Historical and current perspectives on stress and health* (Vol. 2). Oxford, UK: Elsevier.

Sørensen, O. H., Hasle, P., Hesselholt, R. R., & Herbøl, K. (2012). *Nordiske Perspektiver på Arbejdsmiljø—Mening, Indflydelse og samarbejde* [*Nordic work environment perspectives—Mening, control and cooperation*]. NMR-rapport, Det Nationale Forskningscenter for Arbejdsmiljø. Nordisk Ministerråds rapportserie.

Tvedt, S. D., Saksvik, P. Ø., & Nytrø, K. (2009). Healthy organizational change processes as a prerequisite for a good psychological work environment. *Work & Stress, 23,* 80–98.

van Vegchel, N., de Jonge, J., Bosma, H., & Schaufeli, W. (2005). Reviewing the effort-reward imbalance model: Drawing up the balance of 45 empirical studies. *Social Science and Medicine, 60*(5), 1117–1131. doi:10.1016/j.socscimed.2004.06.043.

Warr, P. (1987). *Work, unemployment and mental health.* New York, NY: Oxford University Press.

Warr, P. (2007). *Work, happiness, and unhappiness*. Mahwah, NJ: Lawrence Erlbaum Associates.

Chapter 4
Facilitating a Meaningful Work Situation—A Double-Edged Sword?

Thomas Christian Espenes and Fay Giæver

For employees in today's society, work fulfils a wide range of functions. Work is not only important in terms of invome, it is also a source of identity, of positive self-image and a sense of meaningfulness (Alvesson, 2001; Noon & Blyton, 2007; Wrzesniewski, LoBuglio, Dutton, & Berg, 2013). In a well-known study, a large number of individuals from different countries were asked the so-called lotto question (MOW—International Research Team, 1987). When being presented with the scenario of winning a large sum of money whereby they would not have to work due to financial needs, an overwhelming number of respondents answered that they would continue working. Another study conducted in America investigated individuals who actually won large lottery sums and found that 86% of the winners still continued working (Arvey, Harpaz, & Liao, 2004).

The fact that working is central to well-being and that it helps to create meaning in people's lives is reflected in well-developed welfare states, such as Norway. Throughout history, the understanding of "meaning" has had an impact on how Norwegian working life has developed, as well as on how work today is organized. In the 1970s, Thorsrud and Emery (1970) established a list of "psychological job demands" with the criteria: co-determination, variation and value of the job that were thought to be a source of meaningful working life. These criteria have, in turn, influenced Norwegian law. A preamble (§ 1-1a) to the Working Environment Act states that the law shall "secure a working environment that provides a basis for a health promotion and for meaningful working situation" (Norwegian Ministry of Labour and Social Affairs, 2005).

It can be argued that experiencing work as meaningful may be triggered by, as well as lead to, increased employee work engagement. Although meaningful work

T.C. Espenes · F. Giæver (✉)
Department of Psychology, Norwegian University of Science and Technology (NTNU), Trondheim, Norway
e-mail: fay.giaver@ntnu.no

M. Christensen et al. (eds.), *The Positive Side of Occupational Health Psychology*,
https://doi.org/10.1007/978-3-319-66781-2_4

has hot been clearly defined (Cartwright & Holmes, 2006; but see next section), the essence of meaning is "connection" (Baumeister & Vohs, 2002). Viewing one's work as meaningful and worthwhile has been linked to a range of positive outcomes including a stronger drive to engage with it (Cartwright & Holmes, 2006). In turn, work engagement has been associated with a range of positive outcomes in the organization (Christian, Garza, & Slaughter, 2011; Salanova, Agut, & Peiró, 2005; Xanthopoulou, Bakker, Demerouti, & Schaufeli, 2009). Many organizations, therefore, conduct different interventions with the intention of promoting work engagement, and in recent times there has been a growth of "artistic" organizational interventions (Darsø, 2006).

If meaning is viewed as "connection", then artistic interventions, such as choir singing at work and the act of directing an orchestra as part of leadership programmes, may be particularly suitable to stimulate engagement and meaning creation among employees, because they aim to trigger not only thought processes, or cognitions, but also emotions (Darsø, 2006, 2009; Simpson & Marshall, 2010). Art processes are also recognized as being particularly powerful when it comes to releasing creativity, innovation and energy (Wennes, 2006). In addition, there is empirical evidence that suggests that participation in cultural activities may promote well-being, engagement and positive health outcomes (e.g. Bygren et al., 2009a, b; Bygren, Konlaan, & Johansson, 1996; Cuypers et al., 2011). Choral singing has, for instance, been found to contribute to a positive environment, social cohesion, and increased well-being both in the workplace and in the community (Giæver, Vaag, & Wennes, 2016; Clift & Hancox, 2010; Purcell & Kagan, 2007; Dunbar, Kaskatis, Macdonald, & Barra, 2012). Such interventions can help to enhance the meaning and experience of work.

In this chapter, we explore how meaningful work can be facilitated. First, we explain what is meant by meaningful work. Second, we examine artistic interventions as an organizational strategy for facilitating and stimulating a meaningful working situation. Finally, we discuss the concept of meaning and if striving for meaning is actually desired, or should be desired, by all employees, and at all times.

4.1 Defining Meaningful Work

For the present purpose, meaningful work is defined as three interrelated dimensions. First, as a positive psychological state whereby people feel they make a positive, important, and useful contribution to a worthwhile purpose through the execution of their work (Albrechts, 2013, p. 239); second, as "connection" (Baumeister & Vohs, 2002), coherence and belonging; third, as an alignment of one's values and purpose or, according to Chalofsky (2003), an "integrated wholeness", which is close to the concept of intrinsic motivation. The extent to which work is perceived as meaningful relies not only on subjective experiences, but also on what the individual does to create these experiences and how the psychosocial work environment or culture of the workplace promotes these experiences. This integrated perspective also links well with the four factors implicit to meaningful work identified by Schnell, Höge and Pollet (2013): direction, significance, belonging and coherence.

Making a contribution links clearly to the concept of direction in that it refers to alignment between the individual and the organization's goals, values and norms (Schnell et al., 2013; Kamp, 2012). For example, when the organization is making the case of having a positive impact on clients and customers (Ravn, 2008) a sense of meaningfulness is facilitated among employees. Furthermore, it is essential for employees to experience their leaders as credible and authentic in order for them to accept and share the organization's goals and values (Schnell et al., 2013).

Significance is also about making a contribution as it refers to the ways in which employees perceive that they can make a difference and have impact in the organization and in society overall, something which implies alignment between specific work tasks and the organization's overarching goals (Schnell et al., 2013). Considering that work is increasingly characterized by autonomy, flexibility, and knowledge-based jobs, it can be argued that more pressure is put on the individual employee when it comes to defining the content of his/her job, and hence also meaningfulness (Ravn, 2008).

Making a contribution can also be linked to belonging, or a sense of attachment to a community or the workplace (Schnell et al., 2013), as this is a two-way process whereby the employee is not only drawn to a certain community/workplace, but also actively contributes to its existence. Employees do, however, experience a greater emotional attachment to their work when the culture and formal structure of the organizations facilitate open communication with headroom for constructive conflicts, participation, common trust, as well as support and respect among the workers and the management (Pircher Verdorfer, Weber, Unterrainer, & Seyr, 2012). These characteristics of the work culture are conceptualized as the socio-moral work environment (Weber, Unterrainer, & Schmid, 2009) and have been shown to bolster experienced meaningful work (Schnell et al., 2013). In addition, it has been shown that there is a positive relationship between meaning in work and having appreciative and trustworthy colleagues, as well as supportive management (May, Gilson, & Harter, 2004).

Finally, experiencing cohesion refers to the experience of alignment between one's self-concept and work role (Schnell et al., 2013), something in which also links to make a contribution as this is something in which the individual employee can actively work towards. Hence, if there is a match between the employee's self-concept and the work role it is also easier for the individual to express ideas, values and inner strengths (Schnell et al., 2013) and to experience meaningfulness (Ravn, 2008; May et al., 2004).

With regards to the subjective component of experiences of meaningfulness, it has been argued that work can be viewed either as a job, a career or a calling (Wrzesniewski, McCauley, Rozin, & Schwartz, 1997). Employees who see work as a job are mostly concerned with making money and with material goods.

Employees who see work as a career, on the other hand, often have a deeper personal motivation related to climbing the career ladder, and thereby gaining a higher social status, more power and an improved sense of self-perception. Finally, employees who see work as a calling are less concerned with money and promotion, but view the work as an essential part of their personal identity and are often concerned about contributing to create a better world. There is evidence that when one views the job as a calling, it can bring great satisfaction to both the job and to life itself (Wrzesniewski et al., 1997). Employees who view their job as a calling also feel more passionately about the job and report stronger emotional ties to their job (Vallerand et al., 2003).

With regards to the active component of making a contribution and enhancing meaningfulness, employees can form and redefine the content of their current job through job crafting (Wrzesniewski & Dutton, 2001). As already described in this book, job crafting can consist of crafting tasks (e.g. when one chooses to focus on some tasks rather than others), crafting relationships (e.g. choosing to avoid contact with some of the co-workers and spending more time with others) and/or cognitive crafting (e.g. looking at the job in a wider perspective).

4.2 Artistic Interventions Fostering Meaning

Arts-based methods are increasingly being adopted in order to achieve change in modern and complex organizations where the notion of meaningful work may be particularly important to employees (Adler, 2008; Koivunen & Wennes, 2011). Artistic processes can also be introduced as an intervention in organizations. There are three main types of interventions: primary, secondary and tertiary. These are aimed at preventing a problem, changing an already existing problem or treating the symptoms (Reynolds, 1997; Richardson & Rothstein, 2008). In addition to these, countervailing interventions (Kelloway, Hurrell, & Day, 2008) are characterized by a more proactive approach, where the purpose is to promote and develop positive experiences in the organization. They are, by nature, not necessarily oriented towards the content of the job, but go beyond what the job is about, i.e. with the aim of creating well-being and unity at the workplace in a more holistic way. The expectation is that such interventions can then lead to meaningful work by strengthening the sense of "connection" (Baumeister & Vohs, 2002) and an "integrated wholeness" (Chalofsky, 2003), belongingness and alignment of values. Research has shown that workplaces that support a sense of relatedness, for example, can support intrinsic motivation as part of self-determination (Karanika-Murray & Michaelides, 2015).

Arts-based methods for strengthening the meaning of work can be utilized mainly as a countervailing intervention. A central dimension concerning artistic interventions is that emotional component is especially prominent—this can strengthen the four factors implicit to meaningful work: direction, significance, belonging and coherence (Schnell et al., 2013). Artistic interventions are, in this sense, well aligned to our multidimensional approach which views connection, belonging and an integrated wholeness as central to meaning of work.

Artistic interventions are characterized by the purpose of activating "the whole human being" (Darsø, 2006, 2009; Simpson & Marshall, 2010) whereby cognitive understanding is assumed to be important, but not enough for a deep and long-lasting change in the individual equivalent to double-loop learning (Schein, 1987). When undergoing double-loop learning underlying assumptions are made visible and questioned. In order to achieve this, change/learning the synergy between emotional, bodily and cognitive activation is essential. An emotional activation could constitute either a distinct feeling or a mood condition. Distinct feelings are defined as relatively intense, short-lived reactions to a given situation, a person or an object (Lazarus, 1999; Lazarus & Cohen-Charash, 2001). Moods, on the other hand, are less intense and do not have a particular object, but are closely related to distinct feelings (Parkinson, Totterdell, Briner, & Reynolds, 1996). One can, for example, experience a long-lasting positive mood after having experienced intense joy. Positive emotional experiences (mood and emotions), such as happiness, satisfaction and hope, often arise as a result of being presented with opportunities, achieving one's goal and living by one's norms and values (Bartunek, Rousseau, Rudolph, & DePalma, 2006). It can, therefore, be argued that experiencing a sense of meaningfulness is highly linked to positive emotional experiences and engagement. Furthermore, that artistic intervention can contribute to engagement and meaningfulness via the triggering of positive emotional experiences. Positive emotions have, for instance, been found to be associated with increased openness (Fredrickson & Branigan, 2005), creativity (Isen, Daubman, & Nowicki, 1987) and pro-social and helping behaviour (George & Brief, 1992). It also appears that workplace "fun" has positive effects on individuals and is positively related to job satisfaction and customer service quality (Karl & Peluchette, 2006).

From an intervention point of view, the fact that emotional conditions are short lasting may be seen as a problem, since the intention is to create a long-lasting change in the organization. Still, there are several mechanisms through which emotional experiences may have long-lasting and more profound effects in an organization following an intervention. For instance, most organizational interventions are group- or team-based, so the moods and emotions of individuals converge through emotional contagion (Hatfield, Cacioppo, & Rapson, 1994). The term 'emotional contagion' refers to the process in which individuals automatically tend to mimic and synchronize other people's expressions, vocalizations, postures and movements. Emotional experiences may have more lasting effects as a result of this process. Furthermore, following the premise that positive emotions typically facilitate approach behaviour or continued action, due to a desire to maintain positive emotional experiences (e.g. Carver & Scheier, 1990), Fredrickson (2001) developed the "broaden-and-build" theory. This theory postulates that, although positive moods and emotions are transient and potentially only associated with short-term behavioural effects, positive emotional experiences have the capacity to expand people's thought–action repertoires. As a result of this process, individuals build and accumulate resources over time and can draw on them in the future. There is some empirical support for this theory. For instance, positive emotional experiences have been found to build individual resilience, and to enhance engagement and well-being on both the individual and the collective levels (Fredrickson & Losada, 2005; Fredrickson, Tugade, Waugh, & Larkin, 2003; Hakanen, Bakker, & Demerouti, 2005).

4.3 Types of Artistic Interventions

There are many different types of artistic interventions, and, according to Darsø (2006, 2009) these can be categorized into four levels. Here the idea is that the more levels that are being activated, the greater the potential there is for a deeper and long-lasting learning and changing process in the organization. On the first level, art is utilized as decoration in the organization (e.g. paintings and sculptures in an open plan office), while, on the second level, employees are introduced to art as entertainment (e.g. music during the lunch break). In both of these cases, one the artistic experience may stimulate cognitive and emotional processes and associations among the employees, especially if the employees are actively participating. The process whereby employees are asked to decorate their offices may, for instance, trigger questions such as "what kind of organization is this?" and "how would we like our workplace to be presented?" On the third level, the artistic process is utilized as a method, and the employees and/or the management participate actively in an actual artistic process with the aim of achieving an outcome in the organization. Here, one example would be directing an orchestra in order to stimulate reflection about the leader role, something in which may in turn lead to changes in behaviour through job crafting and an increased sense of meaningfulness (Wrzesniewski & Dutton, 2001; Wrzesniewski et al., 1997; Schnell et al., 2013). The fourth level uses artistic processes as part of an overarching and long-term strategic process in the organization. This implies that the organization invests adequately and that top management believes in the effect of the intervention and express this clearly to employees, such as being expressed by this manager:

> At the end of the day I'm a hard-nosed businessman who wants to sell more washing powder. This is not a soft issue, it's a very hard issue of how you can motivate and inspire people… if I didn't think this program was pulling its weight I would cut it in a second (Darsø, 2009, p. 157).

4.4 Striving After Meaning: A Double-Edged Sword

There is wide range of evidence that introducing artistic interventions in the organization in order to foster positive emotions, engagement and a sense of meaningfulness. It can, however, be questioned whether the stimulation of meaning is always pragmatic or desirable. Could it be possible, that, for example, some employees, at different times in their life, simply wish for a "meaningless" job?

Meaningfulness at work is about making a contribution and links to the experiences of cohesion, direction, significance and belonging (Schnell et al., 2013). The individual employee can also actively contribute to create a meaningful work situation (Wrzesniewski et al., 1997; Wrzesniewski & Dutton, 2001). However, meaning can also be created outside of work, and the experience of meaning can vary in different situations, at different points in one's life, and the experience of meaning in the job can develop over time. In the MOW project (MOW—International Research Team), meaning was explained through a model including six dimensions (Schnell et al., 2013). Here, it is worth highlighting the concept of "work centrality", which refers to how important one's job is at a given time in one's life. Hence, work can be experienced as less essential at various points in the life span, e.g. in the event of working part time due to caring for young children, however, this does not necessarily imply that the job is perceived as meaningless.

Studies have also shown that in the event of involuntary unemployment, meaningful recreational activities can compensate for the satisfaction that work would give (Waters & Moore, 2002). It may even seem as if historical, demographical, job and organizational factors, to a large extent, contribute to perceived job centrality. For example, it seems as if people in our post-industrial society generally value interesting and meaningful work, more than they value materialistic goods (Harding & Hikspoors, 1995). Additionally, several studies show that young employees are more concerned about seeking a high salary, job security and promotion opportunities, compared to older employees who are more concerned with inner rewards (Inceoglu, 2012; Smola & Sutton, 2002). This, however, varies across professions, as nurses often tend to be altruistic and interpersonally oriented in their job (De Cooman et al., 2008) and thereby see their job as a calling and as part of something greater, while industrial workers are more instrumentally oriented (Furåker & Hedenus, 2009). Furthermore, studies have found that those employees born between 1982 and 1999, who constitute the "Generation Me", value spare time more than work, and, compared to earlier generations, wish to work slower and less (Twenge, Campbell, Hoffman, & Lance, 2010). It may, therefore, seem as if work is not crucial for experiencing meaning for this generation.

The centrality and experience of meaning is also something that can be developed over time. The "Generation Me", for instance, grew up in a situation whereby they to a large extent adapted to taking financial security for granted. Hence, they were allowed to immerse themselves in other activities (hobbies etc.), something in which could contribute to viewing work as less meaningful as internal motivation relies on external motivation (Ryan & Deci, 2000). For example, in the event of being motivated by salary and bonuses, the efforts invested in achieving goals could contribute to developing new interests and skills, something which over time can trigger motivation to further pursue new interests and skills. This process can contribute to shifting focus to internal motivation and a sense of meaningfulness, as well as different job crafting activities (Wrzesniewski & Dutton, 2001; Wrzesniewski et al., 1997), that may lead to an enhanced sense of meaning and wider perspectives on work. However, it can also be argued that this process evolves in the opposite direction whereby an employee who is initially altruistically motivated and views the job as a calling, over time can experience an increasing meaninglessness as a result of exhaustion, depersonalization and burnout (Schaufeli, Leiter, & Maslach, 2008).

These processes are often triggered by organizational structures and efficiency demands. In a recent study (Bunderson & Thompson, 2009), it was, for instance, evident that although zoo keepers experienced a strong sense moral commitment and meaningfulness in the job, their spare time, economic situation and social status were negatively affected. For instance, their strong commitment to the job led them to being easily pressured and exploited at work. Hence, experiencing work as deeply meaningful can have negative consequences for the individual employee and imply a destructive dynamic at work, particularly when the goals and demands of the organization (e.g. efficiency) are opposed to the goals and demands of the individual employee (e.g. quality). A similar pattern was evident in a study of sickness presence among Norwegian hospital physicians (Giæver, Lohmann-Lafrenz, & Løvseth, 2016). Presenteeism among physicians has, for example, been linked to more mistakes, more complaints from patients and slower recovery of patients (Virtanen et al., 2009). As expected, we found that the reason the physicians still went to work while sick was largely due to structural conditions. For example, having expertise that was difficult to replace or experiencing job insecurity because they worked in temporary positions. It was evident that the physicians' mindset and compassionate behaviour contributed to maintaining structural limitations. Within this context, presenteeism was considered something positive and as having positive consequences for the physicians. Presenteeism was, for example, viewed as a strategy to handle sickness and as a source of positive professional identity and sense of dignity in their work role. This illustrates that striving for meaning could have long-term negative effects.

4.5 Concluding Thoughts: The Actors of Meaning Creation

The concept of meaning stands strong in Norwegian working life and experienced meaningfulness at work appears to be important for most people. However, considering that the experience of meaning is contextual and evolves in positive as well as negative ways over time, it would not be sensible to define meaning at a general level and then implement prescribed thoughts, ideas and behaviours in the organization. It is essential that employees themselves actively contribute to meaning creation in the organization:

> If we want people's intelligence and support, we must welcome them as co-creators. People support only what they create (Wheatley, 2001 in Darsø, 2006, p. 13)

Here the use of art, or artistic interventions, would be especially well fitted in starting this process, since the aim is to create reflection among employees and where they are active contributors to the intervention process. However, although artistic interventions should be a bottom-up process, it is still important that the management of the organization is actively involved to ensure that individual, social and organizational needs are in balance. For example, management can impose limitations on job crafting so that is not disruptive for the workers or the organization. Management should also be open to challenging organizational goals, and how to approach these in the best possible way.

References

Adler, N. J. (2008). The arts & leadership: Now that we can do anything, what will we do? *Academy of Management Learning & Education, 54*(4), 486–499. doi:10.5465/AMLE.2006.23473209.

Albrecht, S. L. (2013). Work engagement and the positive power of meaningful work. In A. B. Bakker (Ed.), *Advances in positive organizational psychology* (Vol. 1, pp. 237–260). Bradford: Emerald Insight. doi:10.1108/S2046-410X(2013)0000001013.

Alvesson, M. (2001). Knowledge work: Ambiguity, image and identity. *Human Relations, 54*(7), 863–886.

Arvey, R. D., Harpaz, I., & Liao, H. (2004). Work centrality and post award work behavior of lottery winners. *Journal of Psychology, 138*(5), 404–420.

Bartunek, J. M., Rousseau, D. M., Rudolph, J. W., & DePalma, J. A. (2006). On the receiving end: sense-making, emotions and assessments of an organizational change initiated by others. *The Journal of Applied Behavioral Science, 42,* 182–206.

Baumeister, R. F., & Vohs, K. D. (2002). The pursuit of meaningfulness in life. In C. R. Snyder & S. J. Lopez (Eds.), *The handbook of positive psychology*. New York: Oxford University Press.

Bunderson, S. J., & Thompson, J. A. (2009). The call of the wild: Zookeepers, callings and the double-edged sword of deeply meaningful work. *Administrative Science Quarterly, 54,* 32–57.

Bygren, L. O., Johansson, S., Konlaan, B. B., Gribovski, A. M., Wilkinson, A. V., & Sjöström, M. (2009a). Attending cultural events and cancer mortality: A Swedish cohort study. *Arts & Health: An International Journal for Research, Policy and Practice, 1*(1), 64–73. doi:10.1080/17533010802528058.

Bygren, L. O., Johansson, S., Konlaan, B. B., Gribovski, A. M., Wilkinson, A. V., & Sjöström, M. (2009b). Cultural participation and health: A randomized controlled trial among medical care staff. *Psychosomatic Medicine, 71*(4), 469–473. doi:10.1097/PSY.0b013e31819e47d4.

Bygren, L. O., Konlaan, B. B., & Johansson, S. (1996). Attendance at cultural events, reading books or periodicals, and making music or singing in a choir as determinants for survival: Swedish interview survey of living conditions. *British Medical Journal, 313,* 1577–1580.

Cartwright, S., & Holmes, N. (2006). The meaning of work: The challenge of regaining employee engagement and reducing cynicism. *Human Resource Management Review, 16*(2), 199–208.

Carver, C. S., & Scheier, M. F. (1990). Origins and functions of positive and negative affect: A control-process view. *Psychological Review, 97*(1), 19.

Chalofsky, N. (2003). An emerging construct of meaningful work. *Human Resource Development International, 6*(1), 69–83.

Christian, M. S., Garza, A. S., & Slaughter, J. E. (2011). Work engagement: A quantitative review and test of its relations with task and contextual performance. *Personell Psychology*, 89–136. doi:10.1111/j.1744-6570.2010.01203.x.

Clift, S., & Hancox, G. (2010). The significans of choral singing for sustaining psychological wellbeing. *Music Performance Research, 3*(1), 79–96.

Cuypers, K. F., Knudtsen, M. S., Sandgren, M., Krokstad, S., Wikström, B. M., & Theorell, T. (2011). Cultural activities and public health: Research in Norway and Sweden. An overview. *Arts & Health, 3*(1), 6–26. doi:10.1080/17533015.2010.481288.

Darsø, L. (2006). Forandring gennem kunst. *Implementering, Børsen Ledelseshåndbøger, 2*(Maj), 1–16. From: http://pure.au.dk/portal/en/publications/forandring-gennemkunst%282e9a1800-9d99-11dc-8d06-000ea68e967b%29.html.

Darsø, L. (2009). *Artful creation. Learning-tales of arts-in-business*. Frederiksberg, Denmark: Samfundslitteratur.

De Cooman, R., De Gieter, S., Pepermans, R., Du Bois, C., Caers, R., & Jegers, M. (2008). Freshmen in nursing: Job motives and work values of a new generation. *Journal of Nursing Management, 16*(1), 56–64.

Dunbar, R., Kaskatis, K., Macdonald, I., & Barra, V. (2012). Performance of music elevates pain threshold and positive affect: Implications for the evolutionary function of music. *Evolutionary Psychology, 10*(4), 688–702.

Fredrickson, B. L. (2001). The role of positive emotions in positive psychology: The broaden-and-build theory of positive emotions. *American Psychologist, 56*(3), 218–226. doi:10.1037//0003-066X.56.3.218.

Fredrickson, B. L., & Branigan, C. (2005). Positive emotions broaden the scope of attention and thought-action repertoires. *Cognition and Emotion, 19*(3), 313–332.

Fredrickson, B. L., & Losada, M. F. (2005). Positive affect and the complex dynamics of human flourishing. *American Psychologist, 60*(7), 678.

Fredrickson, B. L., Tugade, M. M., Waugh, C. E., & Larkin, G. R. (2003). What good are positive emotions in crisis? A prospective study of resilience and emotions following the terrorist attacks on the United States on September 11th, 2001. *Journal of Personality and Social Psychology, 84*(2), 365.

Furåker, B., & Hedenus, A. (2009). Gammbling windfall decisions: Lottery winners and employment behavior. *UNLV Gaming Research & Review Journal, 13*(2), 1–15.

George, J. M., & Brief, A. P. (1992). Feeling good-doing good: A conceptual analysis of the mood at workorganizational spontaneity relationship. *Psychological Bulletin, 112*(2), 310.

Giæver, F., Lohmann-Lafrenz, S., & Løvseth, L. T. (2016a). Why hospital physicians attend work while ill? The spiralling effect of positive and negative factors. *BMC Health Services Research, 16*, 584.

Giæver, F., Vaag, J., & Wennes, G. (2016). Choral singing as an arts-based organisational intervention: A qualitative study of employees' experiences. *Arts and Health, 9*.

Hakanen, J. J., Bakker, A. B., & Demerouti, E. (2005). How dentists cope with their job demands and stay engaged: The moderating role of job resources. *European Journal of Oral Sciences, 113*(6), 479–487.

Harding, S. H., & Hikspoors, F. J. (1995). New work values: In theory and practice. *International Social Science Journal, 47*, 441–455.

Hatfield, E., Cacioppo, J. T., & Rapson, R. L. (1994). *Emotional contagion. Studies in emotion & social interaction*. Cambridge University Press.

Inceoglu, I. (2012). Age-related differences in work motivation. *Journal of Occupational Psychology, 86*, 300–329.

Isen, A. M., Daubman, K. A., & Nowicki, G. P. (1987). Positive affect facilitates creative problem solving. *Journal of Personality and Social Psychology, 52*(6), 1122.

Kamp, A. (2012). The quest for the meaning of work: Competing concepts of meaning. In A. Kamp & H. Hvid (Eds.), *Elderly care in transition—Management, meaning and identity at work. A Scandinavian perspective* (pp. 51–82). Copenhagen: Copenhagen Business School Press.

Karanika-Murray, M., & Michaelides, G. (2015). Workplace design: Conceptualizing and measuring workplace characteristics for motivation. *Journal of Organizational Effectiveness: People and Performance, 2*(3), 224–243.

Karl, K., & Peluchette, J. (2006). How does workplace fun impact employee perceptions of customer service quality? *Journal of Leadership & Organizational Studies, 13*(2), 2–13. doi:10.1177/10717919070130020201.

Kelloway, E. K., Hurrell, J. J., Jr., & Day, A. (2008). 20 Workplace interventions for occupational stress. In *The individual in the changing working life* (p. 419).

Koivunen, N., & Wennes, G. (2011). Show us the sound! Aesthetic leadership of symphony orchestra conductors. *Leadership, 7*(1), 51–71. doi:10.1177/1742715010386865.

Lazarus, R. S. (1999). *Stress and emotion: A new synthesis*. UK: Springer Publishing Company Inc.

Lazarus, R. S., & Cohen-Charash, Y. (2001). Discrete emotions in organizational life. In R. L. Payne & C. L. Cooper (Eds.), *Emotions at work* (pp. 157–172). New York: Guildford Press.

May, D. R., Gilson, R. L., & Harter, L. M. (2004). The psychological conditions of meaningfulness, safety and availability and the engagement of the human spirit at work. *Journal of Occupational and Organizational Psychology, 77*(1), 11–37. doi:10.1348/096317904322915892.

MOW—International Research Team. (1987). *The meaning of working*. London: Academic Press.

Noon, M., & Blyton, P. (2007). *The realities of work: Experiencing work and employment in contemporary society* (3rd ed.). New York: Palgrave Macmillan.

Norwegian Ministry of Labour and Social Affairs. (2005). *Arbeidsmiljøloven.* Retrieved from, Lovdata.no: http://lovdata.no/dokument/NL/lov/2005-06-17-62.

Parkinson, B., Totterdell, P., Briner, R. B., & Reynolds, S. (1996). *Changing moods: The psychology of mood and mood regulation.* London: Longman.

Pircher Verdorfer, A., Weber, W. G., Unterrainer, C., & Seyr, S. (2012). The relationship between organizational democracy and socio-moral climate: Exploring effects of the ethical context in organizations. *Economic and Industrial Democracy*, 1–27. doi:10.1177/0143831X12450054.

Purcell, C., & Kagan, C. (2007). *Joy at work: The impact of non-professional singing workshops on employee wellbeing.* Report of the Joy at Work Singing Workshops included as part of the Research Institute for Health and Social Change Annual Conference, Manchester.

Ravn, I. (2008). Mening i arbejdslivet - definition og konceptualisering. [Meaning in work—Definition and conceptualization] *Tidsskrift for Arbejdsliv, 10*(4), 59–75.

Reynolds, S. (1997). Psychological well-being at work: Is prevention better than cure? *Journal of Psychosomatic Research, 43*(1), 93–102.

Richardson, K. M., & Rothstein, H. R. (2008). Effects of occupational stress management intervention programs: A meta-analysis. *Journal of Occupational Health Psychology, 13*(1), 69–93.

Ryan, R. M., & Deci, E. L. (2000). Intrinsic and extrinsic motivations: Classic definitions and new directions. *Contemporart Educational Psychology, 25*(1), 54–67.

Salanova, M., Agut, S., & Peiró, J. M. (2005). Linking organizational resources and work engagement to employee performance and customer loyalty: The mediation of service climate. *Journal of Applied Psychology*, 1217–1227. doi:10.1037/0021-9010.90.6.1217.

Schaufeli, W. B., Leiter, M. P., & Maslach, C. (2008). Burnout: 35 years of research and practice. *Career Development International, 14*(3), 204–220.

Schein, E. (1987). *Organizational culture and leadership* (2nd ed.). San Francisco: Jossey Bass.

Schnell, T., Höge, T., & Pollet, E. (2013). Predicting meaning in work: Theory, data, implications. *The Journal of Positive Psychology: Dedicated to furthering research and promoting good practice, 8*(6), 543–554. doi:10.1080/17439760.2013.830763.

Simpson, B., & Marshall, N. (2010). The mutuality of emotions and learning in organizations. *Journal of Management Inquiry, 19*(4), 351–365. doi:10.1177/1056492610376531.

Smola, K. W., & Sutton, C. D. (2002). Generational differences: Revisiting generational work values for the new millennium. *Journal of Organizational Behavior, 23,* 363–382.

Thorsrud, E., & Emery, F. (1970). *Mot en ny bedriftsorganisasjon.* Oslo: Tanum.

Twenge, J. M., Campbell, S. M., Hoffman, B. J., & Lance, C. E. (2010). Generational differences in work values: Leisure and extrinsic values increasing, social and intrinsic values decreasing. *Journal of Management, 36,* 1117–1142.

Vallerand, R. J., Blanchard, C., Mageau, G. A., Koestner, R., Ratelle, C., Leonard, M., et al. (2003). On obsessive and harmonious passion. *Journal of Personality & Social Psychology, 85* (4), 756–767.

Virtanen, M., Kurvinen, T., Terho, K., Oksanen, T., Peltonen, R., Vahtera, J., et al. (2009). Work hours, work stress, and collaboration among ward staff in relation to risk of hospital-associated infection among patients. *Medical Care, 47*(3), 310–318. doi:10.1097/MLR.0b013e3181893c64.

Waters, L. E., & Moore, K. A. (2002). Reducing latent deprivation during unemployment: The role of meaningful leisure activity. *Journal of Occupational and Organizational Psychology, 75*(1), 15–32. doi:10.1348/096317902167621.

Weber, W. G., Unterrainer, C., & Schmid, B. E. (2009). The influence of organizational democracy on employees' socio-moral climate and prosocial behavioral orientation. *Journal of Organizational Behavior, 30*(8), 1127–1149. doi:10.1002/job.615.

Wennes, G. (2006). *Kunstledelse, om ledelse av og i kunstneriske virksomheter*. Oslo: Abstrakt forlag.

Wrzesniewski, A., & Dutton, J. (2001). Crafting a job: Revisioning employees as active crafters of their work. *Academy of Management Review, 26,* 179–201.

Wrzesniewski, A., LoBuglio, N., Dutton, J. E., & Berg, J. M. (2013). Job crafting and cultivating positive meaning and identity in work. *Advances in Positive Organizational Psychology, 1,* 281–302.

Wrzesniewski, A., McCauley, C., Rozin, P., & Schwartz, B. (1997). Jobs, careers, and callings: People's relations to their work. *Journal of Research in Personality, 31,* 21–33.

Xanthopoulou, D., Bakker, A. B., Demerouti, E., & Schaufeli, W. B. (2009). Work engagement and financial returns: A diary study on the role of job and personal resources. *Journal of Occupational and Organizational Psychology, 82,* 183–200. doi:10.1348/096317908X285633.

Chapter 5
Authentic Leadership, Psychological Capital, and Employees' Well-Being

Oyeniyi Samuel Olaniyan

John C. Maxwell, a critically acclaimed American expert on leadership, once said, "everything rises and falls on leadership" (Maxwell, 2007, p. 267). Taken literally, this means that simple organizational variables, like a successful psychosocial work environment, employee satisfaction, absenteeism and presenteeism, job insecurity, and intention to quit the organization, to name a few, could be assigned to the leadership's efforts or lack thereof. More concretely, should the leadership of any organization be held accountable for any result or development both with respect to the employees and to the organization as a whole? Is it logical for us to accept Maxwell's proposal hook, line, and sinker?

Today, we are witnessing an upsurge in various types of leadership theories. From transformational to ethical, servant, laissez faire, destructive, and health-promoting leadership theories among others, it appears that research on leadership has consistently led to the development of several more theories and models (Antonakis, 2003; Avolio & Bass, 2002; Avolio, Bass, & Jung, 1999; Avolio & Gardner, 2005; Barling, Weber, & Kelloway, 1996; Bass & Avolio, 1990; Einarsen, Aasland, & Skogstad, 2007; Gardner, Avolio, Luthans, May, & Walumbwa, 2005; Howell & Avolio, 1993; Howell, Avolio, & Schmitt, 1993; Kelloway, Mullen, & Francis, 2006; Skogstad, Einarsen, Torsheim, Aasland, & Hetland, 2007; Yukl, 2008). Apart from the leadership theories presented above, there is a handful of recently proposed theories focusing more on the direct influence of a leader on subordinates' behavior and well-being, as well as on the many roles of leadership in an organization (Avolio, Gardner, Walumbwa, Luthans, & May, 2004; Eid, Mearns, Larsson, Laberg, & Johnsen, 2012; Peus, Wesche, Streicher, Braun, & Frey, 2012; Yammarino, Dionne, Schriesheim, & Dansereau, 2008). One such theory is authentic leadership style with its origin in the domain of

O.S. Olaniyan (✉)
Norwegian School of Hotel Management, University of Stavanger,
Ellen & Axel Lunds Hus (EA), Kjell Arholms Gate, 39, 4021 Stavanger, Norway
e-mail: gnn25@hotmail.com; samuel.olaniyan@uis.no

M. Christensen et al. (eds.), *The Positive Side of Occupational Health Psychology*,
https://doi.org/10.1007/978-3-319-66781-2_5

positive psychology and positive organizational behavior (Cameron, Dutton, Quinn, & Caldwell, 2004; Cameron & Spreitzer, 2012; Luthans, 2002; Luthans, & Avolio, 2003; Seligman, 2005; Snyder & Lopez, 2005). Not surprisingly, one might start to question the need for or essence of another leadership theory. In their recent meta-analytic review of both the authentic and transformational leadership constructs for instance, Banks, McCauley, Gardner, and Guler (2016), found, among other things, that the two constructs are not the same as some had previously suggested. Furthermore, authentic leadership influences group and organizational performance and citizen behavior more than the transformational leadership (Banks et al., 2016). As its name implies, authentic leadership is described as the sense of personal ownership of one's "experiences, be they thoughts, emotions, needs, preferences, or beliefs, processes captured by the injunction to oneself," as well as behaviors that are consistent with the true self (Harter, 2005, p. 382). An authentic leader, possessing the four components of self-awareness, relational transparency, internalization of ethical perspective, and balanced processing (these four components will be described later), is able to influence positive outcomes from her subordinates to a very great extent. The fact that an authentic leader is able to positively influence followers' psychological capital (PsyCap–hope, efficacy, resilience, and optimism) makes this construct one of the most valuable leadership theories in the literature. What also makes authentic leadership distinctive is the claim by its proponents that it could always be employed side by side with other notable leadership theories (as will be further explained later).

In the recent past, several scholars (Harter, 1996; Snyder, Rand, & Sigmon, 2005) have for instance shown that adolescents who demonstrated behaviors regarded as authentic were more likely to have greater hope for the future, positive affect, and higher self-esteem compared to their peers whose behaviors were not true to the self. Similarly, the realness, and moral and ethical genuineness of an authentic leader stimulate subordinates' willingness to follow the leader's steps and ways of handling situations (Shamir & Eilam, 2005; Walumbwa, Avolio, Gardner, Wernsing, & Peterson, 2008). Put simply, when a leader demonstrates good judgment, respect for the laws, rules, conduct and ethics of the organization, proponents of the authentic leadership style hold that such a leader becomes a model and watchword for his subordinates. This chapter will take a critical look at the concept of authentic leadership style and its relationship with employees' well-being. In doing that, the chapter will focus first on the relationship between this leadership style and PsyCap, which has been regarded as an important concept when discussing employees' well-being.

5.1 Authentic Leadership

According to Walumbwa et al. (2008), attention paid to the authentic leadership theory has been on the increase in the past decades in connection with positive organizational behavior, ethics, and leadership and positive organizational scholarship literatures. Some researchers maintain that the focus and attention on authentic leadership as a construct has been on the increase in the recent past owing to the ability of authentic leaders to deal with issues far beyond the ordinary (Avolio et al., 2004). Furthermore, the fact that authentic leaders are not only true to themselves, but to everyone they come across, makes such leaders necessary beings in dealing with the multifaceted issues faced by organizations and our modern-day complex societies in general (Avolio et al., 2004). I will describe authenticity first before going deeper into the authentic leadership style.

The Oxford Advanced Learner's Dictionary describes "authentic" as someone or something that is regarded as true, genuine, real, and accurate (Hornby & Cowie, 1989). It has often been found that scholars can be far from the truth whenever they "discover" or come across a novel idea in research (Novicevic, Harvey, Ronald, & Brown-Radford, 2006). Like any other new discovery or novel idea within social research, authentic leadership has been around since long before the twenty-first century (Novicevic et al., 2006). Authenticity as a construct has been said to share a lot in common with the commands of ancient Greek philosophers who persuaded Greeks to strive to know themselves and to be true to themselves (Harter, 2005). So the importance of being authentic revolves around getting to know oneself, accepting the result, and remaining true to that self. Barnard (1939) first proposed the idea of an authentic leader who is capable of balancing both her private and public duties in management and organizational studies. As pointed out by some scholars (Avolio et al., 2004), it does one much good to perceive authenticity as a theory existing on a continuum. With that in mind, as the dictionaries also do, scholars have been exercising caution with respect to how they refer to people as authentic or inauthentic, and instead refer to people as more or less authentic. In support of the above, researchers (Avolio et al., 2004; Erickson, 1995; Heidegger, 1962) now share a greater consensus that people become more authentic the truer they are to their fundamental values, and ideas they prefer, including their emotions and identities. So being authentic is never a black-and-white issue.

Authenticity has also been found to be related to self-esteem (Kernis, 2003; Spence Laschinger & Fida, 2014). Kernis (2003) and Spence Laschinger and Fida (2014) maintain that individuals who are able to come to terms with who they are vis-à-vis their strengths and weaknesses are more able to reach levels of stable and high self-esteem compared with those who find it difficult to accept their real self. Individuals who are unable to accept themselves are more prone to defensive biases in contrast to authentic individuals who demonstrate more maturity through their ability to form and maintain stable relationships with others (Spence Laschinger & Fida, 2014). Furthermore, authenticity has been found to have a base in internal self-regulations, that is, authentic individuals are more likely to act according to their internal motives and values rather than external forces, social expectations, incentives, and rewards (Deci & Ryan, 2008).

The definition of authentic leadership style as a construct is built upon many underlying dimensions (Walumbwa et al., 2008). Luthans and Avolio (2003) were the first to define authentic leadership as the "process that draws from both positive psychological capacities and a highly developed organizational context, which results in both greater self-awareness and self-regulated positive behaviors on the part of leaders and associates, fostering self-development" (p. 243). So an authentic leader not only feeds on psychological abilities in order to reach positive results, but she also has a highly developed view of organizational contexts, all of which result in self-development both for the leader and her associates. Apart from the definition of authentic leadership given above, there have been attempts by several other researchers to give their own definition of authentic leadership style. For instance, Avolio, Walumbwa, and Weber (2009) described authentic leadership style as a sequence "of transparent and ethical leader behavior that encourages openness in sharing information needed to make decisions while accepting input from those who follow" (p. 424). Through this definition, it is apparent that the actions of an authentic leader influence her subordinates. Avolio, Luthans, and Walumbwa (2004, as cited in Avolio et al., 2004) also give a more comprehensive definition of authentic leaders in their work on authentic leadership. These researchers maintain that authentic leaders are those who display a high level of awareness in "how they think and behave and are perceived by others as being aware of their own and others' values/moral perspectives, knowledge, and strengths; aware of the context in which they operate; and who are confident, hopeful, optimistic, resilient, and of high moral character" (Avolio, Luthans, & Walumbwa, 2004, p. 4, as cited in Avolio et al., 2004).

5.2 Are Authentic Leaders Born or Made?

Judging by the apparent relevance of all the qualities found in an authentic leader, questions should be asked concerning the emergence, development, and acquisition of these qualities. Specifically, are authentic leaders born or made? According to Northouse (2010), there are two schools of thought within the leadership literature—the process leadership and the trait leadership. Whereas the trait leadership school of thought sees leadership as a personal thing, the process school of thought sees it as interactive, depending largely on the interaction between a leader and her followers. The description of leadership as trait based involves the belief that people are born with some exceptional skills, behaviors, and properties that separate them from others. In his book on leadership, Northouse (2010) maintains that our perception of leadership resides in the definition we give to it. For instance, defining leadership as a "process" removes it from the realms of personality, making leadership a transactional concept occurring between a leader and her followers. The key point here is that influence goes both ways between the leader and her followers (Einarsen & Skogstad, 2015). The idea that leadership is a process makes the concept more available to all and sundry. It also makes it a possibility for those willing to learn and achieve. Some researchers (Avolio, Griffith, Wernsing, & Walumbwa, 2013; Avolio & Hannah, 2008; Garcea, Harrington, & Linley, 2013) maintain that people are more likely to become true leaders as a result of life experiences, rather than heritability.

Given the positive qualities that have been found to be associated with authentic leadership, proper attention ought to be paid to its components. For this reason, the components of authentic leadership style are described below.

5.3 Components of Authentic Leadership

According to Avolio et al. (2013), authentic leadership is comprised of several essential components. These components, briefly described below, are the leader's self-awareness, trigger events, self-reflection, developmental readiness, and core psychological resources. With regard to self-awareness, the leader is expected to demonstrate a higher understanding of herself and the effects of her perceptions on the assessments of situations and people (Avolio et al., 2013). Self-awareness entails the ability to learn one's strong points and weaknesses, while simultaneously developing avenues to reduce the negatives and improve the positives (Eriksen, 2009). We live in a complex world where yesterday's ideas and strategies are becoming obsolete fast; a self-aware leader is crucial in terms of dealing with new situations and challenges in order to guard against poor performance (Avolio et al., 2013).

Trigger events are said to "operate as a form of surprising feedback from other people, major life events, or a perceived success/failure" (Avolio et al., 2013, p. 41). One way through which trigger events could occur is during a leader's interaction with others. Other ways are via training sessions or mere self-reflection. According to Getliffe (1996), self-reflection occurs when a leader consciously and deliberately processes experiences with the aim of gaining knowledge from them. This process is never an automatic one, but occurs hand in hand with experiences and with a sure resolution (Avolio et al., 2013). Self-reflection as one of the components of authentic leadership has been found to occur along a continuum that ranges from positive to negative, that is, from adaptive to maladaptive. Adaptive self-reflection is a process born out of constructive thinking and emotions (Trapnell & Campbell, 1999). Whereas the adaptive process is linked to openness of heart, positive outlook toward happenings, and learning perspectives, maladaptive self-reflection is characterized by damaging ways of thinking that eventually zap energy and performance through engagement in self-doubt, anxiety, and fear-based behaviors (Trapnell & Campbell, 1999).

Like self-reflection, developmental readiness is one of the components of authentic leadership that has been receiving attention lately (Avolio & Hannah, 2008; Hannah & Lester, 2009). In their article on accelerating leader development, Avolio and Hannah (2008) raised the issue concerning the absence of a certified theory and framework for leader development. Furthermore, the scholars claim that techniques for identifying leaders who are developmentally equipped to engage in leader development are still non-existent in the literature. According to Avolio et al. (2013), developmental readiness is born out of a leader's ability, thoughtfulness, and motivational alignment toward the developmental facets of her environment.

Developmental readiness is composed of variables like goal orientation and impetus to leadership development. The level of a leader's developmental readiness is put to test whenever she is faced with trigger events. Avolio et al. (2013) maintain that learning opportunities and a chance to improve the self is often the result for those leaders high in developmental readiness whenever they are confronted with trigger events. On the contrary, leaders who are low in developmental readiness are more likely to dive into a great deal of discounting and defensiveness in similar situations. This defensiveness and ego (self) protection often found with leaders low in developmental readiness may hinder these leaders from learning new ways of doing things. The same cannot be said of their counterparts who are high in developmental readiness, as these leaders welcome trigger events and treat such events as learning opportunities, resulting in more experience and positive learning. No wonder Avolio et al. (2013) called leaders high in developmental readiness the "instigators of such triggers and recipients of their effects" (p. 42).

The last component of authentic leadership is the core positive psychological resources. There are assertions that the core positive psychological resources component of an authentic leader is enjoying more research attention than the remaining components (Avolio & Gardner, 2005). Core positive psychological resources are found in a leader's ability to possess a positive outlook toward who she is, as well as what she can achieve through others (Avolio & Gardner, 2005). True self-reflection and positive change in behavior become the results when a leader's focus is on the demonstration of constant positive leadership behavior (Luthans & Avolio, 2003). As postulated by Avolio et al. (2013), the positive psychological resources as well as two of the aforementioned components (advancement of an adaptive self-reflection and self-awareness) together represent authentic leadership development. The measurement of authentic leadership is described below.

5.4 Measuring Authentic Leadership

The initial intention of the proponents of authentic leadership was to create and develop a construct that is not only multidimensional, but also multilevel (Luthans & Avolio, 2003). This intention, although well meant (Avolio et al., 2013), has received quite some reactions from scholars (Cooper, Scandura, & Schriesheim, 2005). In their work on potential challenges to developing authentic leadership theory and authentic leaders, Cooper et al. (2005) stress the importance of a clear definition of authentic leadership as a construct before embarking on its measurement. Subsequently, this has resulted in a more refined and revised construct (Avolio et al., 2013). Some researchers (Ilies, Morgeson, & Nahrgang, 2005) suggested a four-component model of authentic leadership based on the earlier works of Kernis (2003). These four components are self-awareness, unbiased processing, authentic behavior (acting), and authentic relational orientation.

In conjunction with the above, Gardner et al. (2005) suggested two components (self-awareness and self-regulation) for authentic leadership. Unlike Ilies et al. (2005), Gardner et al. argue that variables like relational transparency, balanced processing of information, internalized regulation, and authentic behavior are associated with the authentic self-regulation process, that is, one of the two aforementioned components of authentic leadership. Up until recently, there have been diverse suggestions from different researchers regarding what and what components make up the authentic leadership construct (Avolio et al., 2013). The current validated instrument designed to measure authentic leadership consists of four dimensions that will be described below.

While ethical behaviors and integrity are among the characteristics that we mostly seek from those in leadership positions (Leroy, Palanski, & Simons, 2012), instruments or means for measuring these qualities are either non-existent or simply lacking proper validation (Avolio et al., 2013). Expecting leaders to show high morals and integrity without any valid psychometric tools for measuring those behaviors will often result in futile efforts (Avolio et al., 2013). After a far-reaching review of the literature on authentic leadership, as well as numerous discussions with prominent intellectuals and researchers on leadership around the world, Walumbwa et al. (2008) first proposed a five-dimensional (self-awareness, balanced information processing, relational transparency, internalized regulatory processing, and moral perspective) construct of authentic leadership. Furthermore, the scholars conducted several interviews involving doctoral students. These interviews gave rise to the merging of the last two dimensions, i.e., internalized regulatory processing and moral perspective. According to Avolio et al. (2013), this merger came into being because of the great similarity between these two dimensions. The final four dimensions of authentic leadership are internalization of an ethical perspective, self-awareness, balanced processing, and relational transparency (Avolio et al., 2013).

5.5 Cultural Generalizations

As part of a wide study on culture, leadership, and organizations, House, Hanges, Javidan, Dorfman, and Gupta (2004) stress the need for the focus on cross-cultural differences. Additionally, in the aftermath of the increase in the number of multicultural organizations today, scholars maintain that "people and organizations from different cultures come into regular contact as customers, competitors, partners, and suppliers. As a result, there is an academic and practical need to understand what happens when cultures connect" (House et al., 2004, p. 730). Additionally, Avolio et al. (2013) maintain that the establishment of higher levels of transparency in the relationship between a leader and her followers (one of the essences of an authentic leader) might pose several challenges in societies where individual thoughts and emotions are seldom celebrated (collectivistic societies) compared to those societies in which importance is placed on such characteristics (individualistic societies).

Although leadership transparency might prove to be a rare commodity in collectivistic societies (owing to the focus on the maintenance of group harmony), the authentic leadership construct in its entirety "reflects a developmental process by which leaders develop themselves and consider the developments of the constituents they lead" (Avolio et al., 2013, p. 48). In line with this thinking, researchers from diverse cultural backgrounds have attempted to investigate the authentic leadership construct and its influence on workplace well-being (Rahimnia & Sharifirad, 2015; Spence Laschinger & Fida, 2014), positive work climate and gender (Woolley, Caza, & Levy, 2011), employees' performance and relational process (Gill & Caza, 2016; Wang, Sui, Luthans, Wang, & Wu, 2014). These researchers have found associations between authentic leadership and all of the above-mentioned variables in their studies.

Through samples collected across different countries (Kenya, China, and the USA), Walumbwa et al. (2008) have been able to validate authentic leadership as a four-dimensional construct in a multicultural setting. Additionally, the researchers found that these four dimensions, working together as a single factor (authentic leadership construct), were positively associated with self-reported measures of organizational commitment and citizen behavior, job satisfaction, and satisfaction with a supervisor. The positive association of the authentic leadership constructs with the aforementioned positive organizational variables was not only found to be significant, but its predictability was much higher than, for example, ethical and transformational forms of leadership (Walumbwa et al., 2008).

In their study on the impact of authentic leadership on employees' creativity, Rego, Sousa, Marques, and Cunha (2012) collected data from about 200 employees from 33 different commerce companies around Portugal. The scholars found that authentic leadership predicted and promoted employees' creativity directly and indirectly through PsyCap. They concluded that authentic leaders not only influence their employees positively, but that these leaders' behaviors promote employees' PsyCap, which in turn helps employees' creativity. Similarly, Borgersen, Hystad, Larsson, and Eid (2014) conducted a study of authentic leadership and the safety climate among seafarers of Filipino origin sailing on various merchant vessels internationally. They found among other things that authentic leadership had a positive and significant association with how workers perceived the safety climate on board vessels. Besides the above-mentioned studies, several other studies (see the introduction) have attempted to validate and verify the influence of the authentic leadership construct on employees' performance and welfare. More recently, Olaniyan and Hystad (2016) reported that authentic leadership has both a direct and indirect effect on job satisfaction, job insecurity, and intention to quit the organization. As I mentioned earlier in the introductory section, an authentic leader's positive influence on her subordinates has made this construct attractive to scholars. This probably explains why many studies (e.g., Borgersen et al., 2014; Latham et al., 2013; Rego et al., 2012) have investigated the role of a leader on the well-being, performance, and health of the employees under them. The aforementioned studies, as well as several others (e.g., Olaniyan & Hystad, 2016), have been able to discover an indirect relationship through PsyCap between authentic leadership and other tangible psychosocial variables such as job satisfaction, employees' performance, employees' creativity, and safety perceptions among other things. Added to these are studies implicating the authentic leader in issues surrounding the safety climate within the shipping industry (Olsen, Larsson, Eid, Borgersen, & Hystad, 2012). It appears that authentic leaders stimulate and activate essential properties or modes of behavior in their employees. These properties in the employees have been labeled as PsyCap and will be presented below.

5.6 What Is the Relationship Between Authentic Leadership Style and PsyCap?

Firstly, both constructs emanate from positive organizational scholarship (POS) and positive organizational behavior (POB). Secondly, in line with the old saying that, "you cannot give what you do not have" (DeVille, 2016), an authentic leader is believed to possess the four components (hope, efficacy, resilience, and optimism) of PsyCap. Although not very clear, the claim is that much of the makeup of PsyCap is found already in an authentic leader's positive psychological resources. Take trigger events as an example; it takes the combination of both resilience and hope for any leader to be able to make anything useful out of complicated and unplanned events (Luthans, Youssef, & Avolio, 2007). Finally, a leader who is able to demonstrate all of the aforementioned components of authentic leadership is able to stimulate hope, efficacy, resilience, and optimism among her employees.

5.7 Psychological Capital

According to Newman, Ucbasaran, Zhu, and Hirst (2014), the proponents of PsyCap efforts were focused on making PsyCap a construct that is state-like as opposed to trait-like. The reason for this is that state-like constructs have the tendency to be less rigid than trait-like constructs. This in turn creates opportunities for learning, thereby meaning that PsyCap is not only restricted to some personality types or characteristics (Dawkins, Martin, Scott, & Sanderson, 2013; Luthans, Avey, Clapp-Smith, & Li, 2008). Emanating from positive psychology and positive organizational behavior, PsyCap focuses more on the positive events connected with a working environment. Specifically, PsyCap came into existence partly as a response to the upsurge in the amount of leaders and scholars who are directing their attention principally to what is wrong within an organization and how to eradicate negative experiences (Luthans et al., 2007a, b).

As they stress in their book on PsyCap, "positive psychology has broadened the perspective beyond what is wrong with people toward optimal functioning, flourishing, and reaching human potential (Luthans et al., 2007a, b, p. vii). Although

Luthans et al. (2007a, b) are not the trailblazers in this line of thought, they maintain that organizations today face bigger challenges than ever before. Not only do these organizations have to engage in competitions for the best human resources and talents, but additionally, it seems that shared commitment and loyalty on the part of both employees and employers is fast becoming a thing of the past (Luthans et al., 2007a, b). It is a certainty that on the one hand, employers nowadays seem to focus more on cutting costs, profitability, and effective management, whereas on the other hand, employees are found to be preoccupied with improving themselves and their CVs. Put simply, the days of retirement with a gold wristwatch are long gone (Saksvik & Nytrø, 2006). Added to these issues is the call for the disappearance of the old order (traditional lifelong employment in the same organization, along with the parental roles of employers over their employees) and the emergence of a new order where employees take responsibility for and ownership of their own (Saksvik & Nytrø, 2006). This particular ownership is referred to as psychological ownership and is said to occur when an individual senses that the mark of ownership of a target or a piece of this target belongs to her (Pierce, Kostova, & Dirks, 2003).

Although psychological ownership for the organization has been found to relate positively with several work environment variables like job satisfaction, self-esteem, and performance, most organizations have been reported to pay increasingly less attention to developing this sort of ownership and the potential of their employees (Avolio, 2005). Furthermore, Luthans et al. (2007a, b) maintain that while the practice of focusing on negative constructs like stress and bullying might serve a purpose, its role has not been found to be sufficient for organizations, especially if these organizations endeavor to sustain a competitive advantage. Additionally, scholars stress that several years of research attention to these negative work environment variables have not been able to further our understanding in the areas of human strengths, ideal functioning, and flourishing (Luthans et al., 2007a, b). Like those before them (e.g., Csikszentmihalyi, 1991; Seligman, 1991), Luthans et al. (2007a, b) promote PsyCap in order to shift our attention from the negative to the positive. Luthans et al. (2007a, b) originally defined PsyCap as:

> An individual's positive psychological state of development… [which] is characterized by: (1) having confidence (self-efficacy) to take on and put in the necessary effort to succeed at challenging tasks; (2) making a positive attribution (optimism) about succeeding now and in the future; (3) persevering toward goals and, when necessary, redirecting paths to goals (hope) in order to succeed; and (4) when beset by problems and adversity, sustaining and bouncing back and even beyond (resiliency) to attain success. (p. 3)

Although researchers argue that PsyCap is composed of the four aforementioned dimensions of self-efficacy, optimism, hope, and resiliency, there are reasons to believe that PsyCap as a construct in its entirety is bigger than all of these four dimensions put together (Luthans et al., 2007a, b). Moreover, these four dimensions seem to interact in more than one way. For example, a hopeful employee might practice hopefulness to the point where she develops resiliency. In a similar fashion, a resilient employee is in a better position to practice adroit and flexible optimism (Luthans et al., 2007a, b).

5.8 Components of PsyCap

According to the proponents of PsyCap, the construct is built on the foundation of the existing concept of positive organizational behavior (Luthans, Youssef, & Avolio, 2015). The four components of PsyCap are presented below.

5.8.1 Self-efficacy

Self-efficacy has been around prior to the development of PsyCap. According to Bandura (1997), self-efficacy is the level of approximation people attribute to completing or achieving a given task or assignment. Self-efficacy does not only motivate and energize one to push toward a goal or an achievement, it also makes an immense contribution to mustering the necessary efforts to achieve this goal (Luthans et al., 2007a, b). Furthermore, self-efficacy does not come on people all of a sudden; it is something that one has both learned and developed over the years (Luthans et al., 2007a, b). PsyCap self-efficacy as Luthans et al. (2007a, b) call it has five essential features. Domain specificity is one of the five essential features of PsyCap self-efficacy that maintains that we are more likely to exhibit uncertainty in some areas of our lives despite our mastery of other areas. Unambiguously, the confidence that we display in one domain might not be freely available in other domains (Luthans et al., 2007a, b).

The second feature of PsyCap self-efficacy is the suggestion that PsyCap is based on mastery or practice. This means that we exhibit mastery in those domains in which we have excelled in the past. As a result, we are more likely to continue our exposure to these domains, which will in turn better our mastery (Luthans et al., 2007a, b). Opposite is the case for the domains about which we are not confident; we might find ourselves avoiding exposure to these domains. This is perhaps related to what was mentioned earlier that we often display confidence in some domains and lack it in others. The third essential feature of PsyCap is related to the first two, and is the idea that there is always room for improvement. Even in the domains in which we have built confidence over the years, we might find that we still need to improve in some areas.

The fourth feature of PsyCap self-efficacy is that our self-efficacy is influenced by others (Luthans et al., 2007a, b). Born from the positive aspect of a self-fulfilling prophecy, the Pygmalion effect stipulates that people perform much better if greater expectation is placed upon them (Luthans et al., 2007a, b). So watching others perform or achieve something creates the confidence in us about that thing, especially if the people are similar to us. The last essential feature of PsyCap self-efficacy is that it is variable. This means that our self-efficacy changes in relation to changes in surrounding factors (Luthans et al., 2007a, b). For instance, an employee might have a genius idea about a product, but the organization in which this employee works might not be able to provide the necessary support for the development of this product (Luthans et al., 2007a, b).

In conjunction with the above, people who are high in self-efficacy are more likely to flourish in challenges, display great self-motivation, put in the necessary work to realize their goals, and will more often than not persevere in the face of obstacles (Luthans et al., 2007a, b). Conclusively, how individuals perceive and interpret events can have a heavy impact on PsyCap efficacy and confidence (Luthans et al., 2007a, b).

5.8.2 Optimism

Optimism, although frequently much talked about, is perhaps the least understood of all the psychological strengths (Luthans et al., 2007a, b). To the average Joe, optimism means looking at the brighter side of life, and pessimism is the opposite of this. PsyCap optimism seems to share the same non-professional's definition of optimism but with a broader perspective (Luthans et al., 2007a, b). According to the exponents of PsyCap optimism, it is not enough to predict the occurrence of good things in the future—the reasons and ascriptions used for describing these occurrences, whether positive or negative, past, present, or future, seem to be of utmost concern. While looking at the brighter side of life is not a bad venture in itself, PsyCap optimism goes further to focus on the mechanisms, motives, and explanations for this decision. As pointed out by Seligman (1991), optimism has a lot to do with an explanatory style of giving positive attributions to events that are personal, permanent, and of ubiquitous origin, as well as to interpreting negative events as external, never long-lasting, and as situation-specific factors. So optimists will more often than not take credit for things working well in their lives (Seligman, 2002). On the contrary, pessimists seldom ascribe to themselves credit for positive occurrences in their lives (Seligman, 1991).

While being optimistic has several advantages, it also has its challenges. It is possible that the ascription of an external explanation to risk factors could present optimistic people (as well as those in their families and organizations) with unanticipated problems (Luthans et al., 2007a, b). When working in an environment characterized by fluctuations and perpetual challenges, it is of great importance to have a mindset that is able to accept errors and be willing to tackle them so they do not occur again. This very fact might become a problem (for an optimist) if one fails to admit the error in the first place. Besides, "explanatory styles are based on one's subjective perceptions and attributions, which may not always be realistic or allow flexibility" (Luthans et al., 2007a, b, p. 95). According to Schneider (2001), maintaining tolerance for the past, gratitude for the present, along with a clear opportunity quest for the future, will help in developing realistic optimism in any given organization.

5.8.3 *Hope*

Like the majority of terms and constructs in psychology, there are misconceptions of what hope is and what constitutes hope as a construct (Luthans et al., 2007a, b). According to Snyder, Irving, and Anderson (1991), hope is "a positive motivational state that is based on an interactively derived sense of successful (1) agency (goal-directed energy) and (2) pathways (planning to meet goals)" (p. 287). Hope is a cognitive state by which an individual through perceived internalized control and self-directed fortitude plus energy can reach out to set goals and face intimidating challenges (Luthans et al., 2007a, b). Two things are important in the concept of hope: will power and pathways. According to the proponents of PsyCap hope, hopeful people are adept at devising new pathways if the original pathways become unnavigable. As one might expect, positive results from creating new paths might give rise to great enthusiasm, as well as to better chances of achieving success.

While granting that hopeful people and organizations have been found to exhibit several positive behaviors and actions (independent thinking, proactivity, etc.), Luthans et al. (2007a, b) warn against unrealistic hope. Even though stretch goals are one of the eight mentioned approaches to developing and nurturing PsyCap hope, having or holding on to unrealistic hope could produce an effect counter to what is intended.

5.8.4 *Resilience*

According to Luthans et al. (2007a, b), resiliency is nothing new for humankind. The world has witnessed it time and again in eminent personalities like Mother Teresa, Winston Churchill, and Nelson Mandela to name a few. The definition of resiliency is part of the general definition of PsyCap, and means the ability to spring back to the surface in the face of imminent problems and challenges. Whenever an employee is facing situations that push her beyond her threshold capacity, such an employee is faced with the opportunity to tap into her own private reserve of resiliency (Luthans et al., 2007a, b). Among other things, research (Luthans, Avey, Avolio, Norman, & Combs, 2006; Luthans, Avolio, Avey, & Norman, 2007) has shown that resiliency is related to job satisfaction and improved performance.

5.9 The Measurement of PsyCap

Luthans et al. (2007a, b) stress the need for a PsyCap construct that is not only reliable and valid, but that is a measure that can undergo development while demonstrating performance impact. In the construction of the PsyCap questionnaire (PCQ), Luthans et al. (2007a, b) drew from already existing constructs like hope, optimism, resilience, and self-efficacy. Since its proposal, PsyCap has been tried and tested by several scholars. The most commonly employed instrument is the four-dimensional 24-item PCQ proposed and validated by Luthans et al. (2007a, b). In their review of PsyCap, Newman et al. (2014) went through sixty different empirical studies and discovered that over half of these studies had employed the 24-item PCQ to capture PsyCap. Some of these empirical studies will be presented below.

5.10 Empirical Evidence

Although the use of the authentic leadership scale in Norway is not as popular as its predecessors, i.e., the transformational and charismatic leadership styles, its usage is currently gaining momentum among researchers, especially within the human factors research area. For instance, Hystad, Bartone, and Eid (2014) recently conducted a study on positive organizational behavior and safety in the offshore oil industry. The study's participants consisted predominantly of Norwegian offshore workers representing a large international petroleum production and exploration company based on the continental shelf of Norway. The researchers found among other things that PsyCap partially mediated the association between authentic leadership and the safety climate. This effect was found to be positive and statistically significant. In a way, findings from Norwegian studies also provide support for the impact of both authentic leadership and PsyCap. In addition to the above-mentioned study, there are several other Norwegian studies of authentic leadership, PsyCap, and employees' well-being (Bergheim, Nielsen, Mearns, & Eid, 2015; Olaniyan & Hystad, 2016).

Similarly, Avey, Luthans, and Youssef (2010) investigated the additive value of positive PsyCap in predicting work attitudes and behaviors. The study consisted of 336 participants selected from a spectrum of organizations and jobs. The scholars hypothesized that PsyCap would have a positive relationship with organizational citizens' behaviors (OCB), and a negative relationship with organizational cynicism, intention to quit the organization, and counterproductive work behaviors (CWB). The results from the study showed that PsyCap was negatively related to organizational cynicism and intention to quit the organization. Additionally, the researchers found that PsyCap was positively related to OCBs. The employees who were high in PsyCap also demonstrated fewer undesirable CWBs.

Apart from the above-mentioned studies, there have been a plethora of studies investigating the essence of PsyCap within organizations (Avey, Reichard, Luthans, & Mhatre, 2011; Avey, Wernsing, & Luthans, 2008; Gooty, Gavin, Johnson, Frazier, & Snow, 2009; Görgens-Ekermans & Herbert, 2013; Jensen & Luthans, 2006; Krasikova, Lester, & Harms, 2015; Luthans, Avolio, Walumbwa, & Li, 2005; Rego et al., 2012; Youssef & Luthans, 2013). This essence of PsyCap within organizations has been found to be especially associated with the presence of an authentic leader.

In concrete terms, when an authentic leader stimulates hope, self-efficacy, resilience, and optimism among her subordinates, there are reasons to believe that these employees more often than not will experience high job satisfaction as well as reduced insecurities and turnover intentions.

5.11 Is Authentic Leadership Style Any Different from Other Leadership Theories?

The proponents of authentic leadership describe the concept as being a "root construct"; this means that the authentic leadership construct is rather generic and could be incorporated with other leadership theories like charismatic, servant, transformational, spiritual, and several other positive leadership styles in the literature (Avolio et al. 2004). Taking self-regulation (one of the components of an authentic leader) as an example, Avolio et al. (2004) claim that being conscious and regulating the self is a vital component whichever leadership theory one is applying. A visionary leader who pushes employees under her toward a goal might suddenly be showing behaviors that are contrary to this vision if self-awareness is not in place (Avolio & Gardner, 2005). On the contrary, an authentic visionary leader will set a vision for her followers while pursuing and making the said vision her own personal priority, not just in words, but also in actions (Avolio & Gardner, 2005).

5.12 Conclusion

Leadership as a construct has had its fair share of focus and attention from scholars in recent years. As mentioned in the introductory section of this chapter, can we actually take Maxwell's (2007) assertion (that everything starts and ends with a leader) seriously, or should we do so with a pinch of salt? As results from the present study have shown, (also in line with past studies of authentic leadership style), findings point to the fundamental significance of having a leader that is perceived by the employees as authentic. Furthermore, evidence for the impact of the authentic leader on her followers through developing employees' PsyCap (i.e., hope, efficacy, resilience, and optimism) was also found. Put together, these findings partially support Maxwell's assertion that a leader's role is a vital one in any given organization.

References

Antonakis, J. (2003). The nature of executive leadership: A conceptual and empirical analysis of success. *Contemporary Psychology-Apa Review of Books, 48*(6), 784–786.

Avey, J. B., Luthans, F., & Youssef, C. M. (2010). The additive value of positive psychological capital in predicting work attitudes and behaviors. *Journal of Management, 36*(2), 430–452. doi:10.1177/0149206308329961.

Avey, J. B., Reichard, R. J., Luthans, F., & Mhatre, K. H. (2011). Meta-analysis of the impact of positive psychological capital on employee attitudes, behaviors, and performance. *Human Resource Development Quarterly, 22*(2), 127–152. doi:10.1002/hrdq.20070.
Avey, J. B., Wernsing, T. S., & Luthans, F. (2008). Can positive employees help positive organizational change? Impact of psychological capital and emotions on relevant attitudes and behaviors. *The Journal of Applied Behavioral Science, 44*(1), 48–70. doi:10.1177/0021886307311470.
Avolio, B. J. (2005). *Leadership development in balance: Made/born*. Mahwah, N.J.: L. Erlbaum.
Avolio, B. J., & Bass, B. M. (2002). *Developing potential across a full range of leadership: Cases on transactional and transformational leadership*. Mahwah, N.J.: Lawrence Erlbaum Associates.
Avolio, B. J., Bass, B. M., & Jung, D. I. (1999). Re-examining the components of transformational and transactional leadership using the multifactor leadership. *Journal of Occupational and Organizational Psychology, 72*(4), 441–462. doi:10.1348/096317999166789.
Avolio, B. J., & Gardner, W. L. (2005). Authentic leadership development: Getting to the root of positive forms of leadership. *The Leadership Quarterly, 16*(3), 315–338. doi:10.1016/j.leaqua.2005.03.001.
Avolio, B. J., Gardner, W. L., Walumbwa, F. O., Luthans, F., & May, D. R. (2004). Unlocking the mask: A look at the process by which authentic leaders impact follower attitudes and behaviors. *The Leadership Quarterly, 15*(6), 801–823. doi:10.1016/j.leaqua.2004.09.003.
Avolio, B. J., Griffith, J., Wernsing, T. S., & Walumbwa, F. O. (2013). What is authentic leadership development? In N. Garcea, S. Harrington, & P. A. Linley (Eds.), *The Oxford handbook of positive psychology and work* (pp. 39–52). Oxford: Oxford University Press.
Avolio, B. J., & Hannah, S. T. (2008). Developmental readiness: Accelerating leader development. *Consulting Psychology Journal: Practice and Research, 60*(4), 331–347. doi:10.1037/1065-9293.60.4.331.
Avolio, B. J., Walumbwa, F. O., & Weber, T. J. (2009). Leadership: Current theories, research, and future directions. *Annual Review of Psychology, 60*(1), 421–449. doi:10.1146/annurev.psych.60.110707.163621.
Bandura, A. (1997). *Self-efficacy: The exercise of control*. New York: Freeman.
Banks, G. C., McCauley, K. D., Gardner, W. L., & Guler, C. E. (2016). A meta-analytic review of authentic and transformational leadership: A test for redundancy. *The Leadership Quarterly, 27*(4), 634–652. doi:10.1016/j.leaqua.2016.02.006.
Barling, J., Weber, T., & Kelloway, E. K. (1996). Effects of transformational leadership training on attitudinal and financial outcomes: A field experiment. *Journal of Applied Psychology, 81*(6), 827–832. doi:10.1037/0021-9010.81.6.827.
Barnard, C. I. (1939). *Dilemmas of leadership in the democratic process*. Princeton: Princeton University Press.
Bass, B. M., & Avolio, B. J. (1990). *Transformational leadership development: Manual for the multifactor leadership questionnaire*. Palo Alto, Calif.: Consulting Psychologists Press.
Bergheim, K., Nielsen, M. B., Mearns, K., & Eid, J. (2015). The relationship between psychological capital, job satisfaction, and safety perceptions in the maritime industry. *Safety Science, 74,* 27–36. doi:10.1016/j.ssci.2014.11.024.
Borgersen, H. C., Hystad, S. W., Larsson, G., & Eid, J. (2014). Authentic leadership and safety climate among seafarers. *Journal of Leadership & Organizational Studies, 21*(4), 394–402. doi:10.1177/1548051813499612.
Cameron, K., Dutton, J., Quinn, R., & Caldwell, C., (2004). *Positive organizational scholarship: Foundations of a new discipline* (Vol. 29, pp. 296–301). Oakland: Berrett-Koehler Publishers.
Cameron, K. S., & Spreitzer, G. M. (2012). *The Oxford handbook of positive organizational scholarship*. Oxford: Oxford University Press.
Cooper, C. D., Scandura, T. A., & Schriesheim, C. A. (2005). Looking forward but learning from our past: Potential challenges to developing authentic leadership theory and authentic leaders. *The Leadership Quarterly, 16*(3), 475–493. doi:10.1016/j.leaqua.2005.03.008.
Csikszentmihalyi, M. (1991). *Flow: Optimaloplevelsens psykologi*. København: Munksgaard.

Dawkins, S., Martin, A., Scott, J., & Sanderson, K. (2013). Building on the positives: A psychometric review and critical analysis of the construct of Psychological Capital. *Journal of Occupational and Organizational Psychology, 86*(3), 348–370. doi:10.1111/joop.12007.

Deci, E. L., & Ryan, R. M. (2008). Self-determination theory: A macrotheory of human motivation, development, and health. *Canadian Psychology/Psychologie Canadienne, 49*(3), 182–185. doi:10.1037/a0012801.

DeVille, M., (2016). Retrieved on 15. November, 2016 from: http://www.evancarmichael.com/library/michele-deville/You-Cant-Give-What-You-Dont-Have.html.

Eid, J., Mearns, K., Larsson, G., Laberg, J. C., & Johnsen, B. H. (2012). Leadership, psychological capital and safety research: Conceptual issues and future research questions. *Safety Science, 50* (1), 55–61. doi:10.1016/j.ssci.2011.07.001.

Einarsen, S., Aasland, M. S., & Skogstad, A. (2007). Destructive leadership behaviour: A definition and conceptual model. *The Leadership Quarterly, 18*(3), 207–216. doi:10.1016/j.leaqua.2007.03.002.

Erickson, R. J. (1995). The importance of authenticity for self and society. *Symbolic Interaction, 18*(2), 121–144.

Eriksen, M. (2009). Authentic leadership: Practical Reflexivity, self-awareness, and self-authorship. *Journal of Management Education*. doi:10.1177/1052562909339307.

Einarsen, S., & Skogstad, A. (2015). *Ledelse på godt og vondt: Effektivitet og trivsel* (2. utg.). Fagbokforlaget 2015 (ISBN 978-82-450-1658-1) 398 s. UiB

Garcea, N., Harrington, S., & Linley, P. A. (2013). *The Oxford handbook of positive psychology and work*. Oxford: Oxford University Press.

Gardner, W. L., Avolio, B. J., Luthans, F., May, D. R., & Walumbwa, F. (2005). "Can you see the real me?" A self-based model of authentic leader and follower development. *The Leadership Quarterly, 16*(3), 343–372. doi:10.1016/j.leaqua.2005.03.003.

Getliffe, K. A. (1996). An examination of the use of reflection in the assessment of practice for undergraduate nursing students. *International Journal of Nursing Studies, 33*(4), 361–374.

Gill, C., & Caza, A. (2016). An investigation of authentic leadership's individual and group influences on follower responses. *Journal of Management*, 014920631456646.

Gooty, J., Gavin, M., Johnson, P. D., Frazier, M. L., & Snow, D. B. (2009). In the Eyes of the Beholder: Transformational Leadership, Positive Psychological Capital, and Performance. *Journal of Leadership & Organizational Studies, 15*(4), 353–367. doi:10.1177/1548051809332021.

Görgens-Ekermans, G., & Herbert, M. (2013). Psychological capital: Internal and external validity of the Psychological Capital Questionnaire (PCQ-24) on a South African sample. *SA Journal of Industrial Psychology, 39*(2), 1-12.

Hannah, S. T., & Lester, P. B. (2009). A multilevel approach to building and leading learning organizations. *The Leadership Quarterly, 20*(1), 34–48. doi:10.1016/j.leaqua.2008.11.003.

Harter, S. (1996). *Historical roots of contemporary issues involving self-concept Handbook of self-concept: Developmental, social, and clinical considerations* (pp. 1–37). Oxford, England: Wiley.

Harter, S. (2005). Authenticity. In C. R. Snyder, & S. J. Lopez, (Eds), *Handbook of Positive psychology* (pp. 382–394). New York: Oxford University Press.

Heidegger, M. (1962). *Being and time* (J. Macquarrie and E. Robinson, Trans.). New York: Harper and Row.

Hornby, A. S., & Cowie, A. P. (1989). *Oxford advanced learner's dictionary of current English* (4th ed. chief editor: A.P. Cowie. ed.). Oxford: Oxford University Press.

House, R. J., Hanges, P. J., Javidan, M., Dorfman, P. W., & Gupta, V. (2004). *Culture, leadership, and organizations: The GLOBE study of 62 societies*. Thousand Oaks, CA: Sage.

Howell, J. M., & Avolio, B. J. (1993). Transformational leadership, transactional leadership, locus of control, and support for innovation: Key predictors of consolidated-business-unit performance. *Journal of Applied Psychology, 78*(6), 891–902. doi:10.1037/0021-9010.78.6.891.

Hystad, S. W., Bartone, P. T., & Eid, J. (2014). Positive organizational behavior and safety in the offshore oil industry: Exploring the determinants of positive safety climate. *The Journal of Positive Psychology, 9*(1), 42–53. doi:10.1080/17439760.2013.831467.

Ilies, R., Morgeson, F. P., & Nahrgang, J. D. (2005). Authentic leadership and eudaemonic well-being: Understanding leader–follower outcomes. *The Leadership Quarterly, 16*(3), 373–394. doi:10.1016/j.leaqua.2005.03.002.

Jensen, S. M., & Luthans, F. (2006). Relationship between entrepreneurs' psychological capital and their authentic leadership. *Journal of Managerial Issues, 18*(2), 254–273. doi:10.2307/40604537.

Kelloway, E. K., Mullen, J., & Francis, L. (2006). Divergent effects of transformational and passive leadership on employee safety. *Journal of Occupational Health Psychology, 11*(1), 76–86. doi:10.1037/1076-8998.11.1.76.

Kernis, M. H. (2003). Toward a conceptualization of optimal self-esteem. *Psychological Inquiry, 14*(1), 1–26. doi:10.1207/S15327965PLI1401_01.

Krasikova, D. V., Lester, P. B., & Harms, P. D. (2015). Effects of psychological capital on mental health and substance abuse. *Journal of Leadership & Organizational Studies*. doi:10.1177/1548051815585853.

Latham, G. P., Piccolo, R., Luthans, F., Youssef, C. M., Sweetman, D. S., & Harms, P. D. (2013). Meeting the leadership challenge of employee well-being through relationship PsyCap and Health PsyCap. *Journal of Leadership & Organizational Studies, 20*(1), 118–133. doi:10.1177/1548051812465893.

Leroy, H., Palanski, M., & Simons, T. (2012). Authentic leadership and behavioral integrity as drivers of follower commitment and performance. *Journal of Business Ethics, 107*(3), 255–264. doi:10.1007/s10551-011-1036-1.

Luthans, F. (2002). The need for and meaning of positive organizational behavior. *Journal of Organizational Behavior, 23*(6), 695–706. doi:10.1002/job.165.

Luthans, F., Avey, J. B., Avolio, B. J., Norman, S. M., & Combs, G. M. (2006). Psychological capital development: Toward a micro-intervention. *Journal of Organizational Behavior, 27*(3), 387–393. doi:10.1002/job.373.

Luthans, F., Avey, J. B., Clapp-Smith, R., & Li, W. (2008). More evidence on the value of Chinese workers' psychological capital: A potentially unlimited competitive resource? *The International Journal of Human Resource Management, 19*(5), 818–827. doi:10.1080/09585190801991194.

Luthans, F., & Avolio, B. J. (2003). Authentic leadership: A positive developmental approach. In K. S. Cameron, J. E. Dutton, & R. E. Quinn (Eds.), *Positive organizational scholarship* (pp. 241–261). San Francisco: Barrett-Koehler.

Luthans, F., Avolio, B. J., Avey, J. B., & Norman, S. M. (2007a). Positive psychological capital: Measurement and relationship with performance and satisfaction. *Personnel Psychology, 60* (3), 541–572. doi:10.1111/j.1744-6570.2007.00083.x.

Luthans, F., Avolio, B. J., Walumbwa, F. O., & Li, W. (2005). The psychological capital of chinese workers: Exploring the relationship with performance. *Management and Organization Review, 1*(2), 249–271. doi:10.1111/j.1740-8784.2005.00011.x.

Luthans, F., Youssef, C. M., & Avolio, B. J. (2007b). *Psychological capital: Developing the human competitive edge*. Oxford: Oxford University Press.

Luthans, F., Youssef, C. M., & Avolio, B. J. (2015). *Psychological capital and beyond*. Oxford; New York: Oxford University Press.

Maxwell, J. C. (2007). *The 21 irrefutable laws of leadership: Follow them and people will follow you*. Nashville: Tennessee, Thomas Nelson Inc.

Newman, A., Ucbasaran, D., Zhu, F., & Hirst, G. (2014). Psychological capital: A review and synthesis. *Journal of Organizational Behavior, 35*(S1), S120–S138. doi:10.1002/job.1916.

Northouse, P. G. (2010). *Leadership: Theory and practice* (5th ed.). Los Angeles: Sage.

Novicevic, M. M., Harvey, M. G., Ronald, M., & Brown-Radford, J. A. (2006). Authentic leadership: A historical perspective. *Journal of Leadership & Organizational Studies, 13*(1), 64–76. doi:10.1177/10717919070130010901.

Olaniyan, O. S., & Hystad, S. W. (2016). Employees' psychological capital, job satisfaction, insecurity, and intentions to quit: The direct and indirect effects of authentic leadership. *Revista de Psicología del Trabajo y de las Organizaciones, 32*(3), 163–171. doi:10.1016/j.rpto.2016.09.003.

Olsen, O. K., Larsson, U. G., Eid, J., Borgersen, H. C., & Hystad, S. W. (2012). Authentic leadership and safety climate in shipping. *International Journal of Psychology, 47*(S1) Suppl. 1s., 541.

Peus, C., Wesche, J., Streicher, B., Braun, S., & Frey, D. (2012). Authentic leadership: An empirical test of its antecedents, consequences, and mediating mechanisms. *Journal of Business Ethics, 107*(3), 331–348. doi:10.1007/s10551-011-1042-3.

Pierce, J. L., Kostova, T., & Dirks, K. T. (2003). The state of psychological ownership: Integrating and extending a century of research. *Review of General Psychology, 7*(1), 84–107. doi:10.1037/1089-2680.7.1.84.

Rahimnia, F., & Sharifirad, M. S. (2015). Authentic leadership and employee well-being: The mediating role of attachment insecurity. *Journal of Business Ethics, 132*(2): 363–377.

Rego, A., Sousa, F., Marques, C., & e Cunha, M. P. (2012). Authentic leadership promoting employees' psychological capital and creativity. *Journal of Business Research, 65*(3), 429–437. doi:10.1016/j.jbusres.2011.10.003.

Saksvik, P. Ø., & Nytrø, K. (2006). *Ny personalpsykologi for et arbeidsliv i endring: nye perspektiver på samspillet organisasjon og menneske* (2. utg. ed.). Oslo: Cappelen akademisk forl.

Schneider, B. (2001). Fits about fit. *Applied Psychology, 50*(1), 141–152. doi:10.1111/1464-0597.00051.

Seligman, M. E. P. (1991). *Learned optimism.* New York: Knopf.

Seligman, M. E. P. (2002). *Authentic happiness: Using the new positive psychology to realize your potential for lasting fulfillment.* New York: Free Press.

Seligman, M. E. P. (2005). Positive psychology, positive prevention, and positive therapy. In C. R. Snyder, & S. J. Lopez, (Eds.), *Handbook of positive psychology* (pp. 3–12). New York: Oxford University Press.

Shamir, B., & Eilam, G. (2005). "What's your story?" A life-stories approach to authentic leadership development. *The Leadership Quarterly, 16*(3), 395–417. doi:10.1016/j.leaqua.2005.03.005.

Skogstad, A., Einarsen, S., Torsheim, T., Aasland, M. S., & Hetland, H. (2007). The destructiveness of laissez-faire leadership behavior. *Journal of Occupational Health Psychology, 12*(1), 80–92. doi:10.1037/1076-8998.12.1.80.

Snyder, C. R., Irving, L., & Anderson, J. (1991). Hope and health. Measuring the will and ways. In C. R. Snyder & D. R. Forsyth (Eds.), *Handbook of social and clinical psychology* (pp. 285–305). Elmsford, NY: Pergamon.

Snyder, C. R., & Lopez, S. J. (2005). *Handbook of positive psychology.* New York: Oxford University Press.

Snyder, C. R., Rand, K. L., & Sigmon, D. R. (2005). Hope theory: Amember of the positive psychology family. In C. R. Snyder, & S. J. Lopez, (Eds), *Handbook of positive psychology* (pp. 257–276). New York: Oxford University Press.

Spence Laschinger, H. K., & Fida, R. (2014). New nurses burnout and workplace wellbeing: The influence of authentic leadership and psychological capital. *Burnout Research, 1*(1), 19–28. doi:10.1016/j.burn.2014.03.002.

Trapnell, P. D., & Campbell, J. D. (1999). Private self-consciousness and the five-factor model of personality: Distinguishing rumination from reflection. *Journal of Personality and Social Psychology, 76*(2), 284–304. doi:10.1037/0022-3514.76.2.284.

Walumbwa, F. O., Avolio, B. J., Gardner, W. L., Wernsing, T. S., & Peterson, S. J. (2008). Authentic leadership: Development and validation of a theory-based measure†. *Journal of Management, 34*(1), 89–126. doi:10.1177/0149206307308913.

Wang, H., Sui, Y., Luthans, F., Wang, D., & Wu, Y. (2014). Impact of authentic leadership on performance: Role of followers' positive psychological capital and relational processes. *Journal of Organizational Behavior, 35*(1), 5–21. doi:10.1002/job.1850.

Woolley, L., Caza, A., & Levy, L. (2011). Authentic leadership and follower development: Psychological capital, positive work climate, and gender. *Journal of Leadership & Organizational Studies, 18*(4), 438–448. doi:10.1177/1548051810382013.

Yammarino, F. J., Dionne, S. D., Schriesheim, C. A., & Dansereau, F. (2008). Authentic leadership and positive organizational behavior: A meso, multi-level perspective. *The Leadership Quarterly, 19*(6), 693–707. doi:10.1016/j.leaqua.2008.09.004.

Youssef, C. M., & Luthans, F. (2013). An integrated model of psychological capital in the workplace. In N. Garcea, S. Harrington, & P. A. Linley (Eds.), *The Oxford handbook of positive psychology and work* (pp. 277–288). Oxford: Oxford University.

Yukl, G. (2008). How leaders influence organizational effectiveness. *The Leadership Quarterly, 19*(6), 708–722. doi:10.1016/j.leaqua.2008.09.008.

Chapter 6
Work, Family, and Leisure

Karoline Grødal

For a lot of people, work and family are the two most dominating spheres in life and the ability to balance them is among the greatest social challenges in our time (Halpern, 2005). This has become an important issue because of several societal changes which have led to new responsibilities for both men and women (Innstrand, Langballe, Falkum, Espnes, & Aasland, 2009). For example, the number of employed women increased drastically in the postwar period (Roos, Trigg, & Hartman, 2006). Today, women constitute 47% of the workforce in Norway and increasingly are working full time (Statistics Norway, 2014). Norwegian women are among those who work most compared with women in other European countries. Concurrently, the traditional nuclear family with the father as provider has become increasingly rare, while dual-career couples, where both are facing demands related to both work and home, are more common. In addition, an increasing number of people are living alone (Statistics Norway, 2014) and are sole providers for children (Statistics Norway, 2013) and therefore presumably have to take greater responsibilities on their own.

In Chap. 1, we briefly discussed the development of the postindustrial society. Jobs are characterized by tasks that are less constricted by time or place, and employees are more often required to be available at all times (Guest, 2002). Information and communication technology have become more significant and have led to changes, not only for the work but also for how we live (Bouwman, van den Hoof, van de Wijngaert, & van Dijk, 2005). Easier access to the Internet, and more sophisticated tools, enables us to constantly stay in touch with our work (Boswell & Olson-Buchanan, 2007) and gives us the opportunity to do non-work-related activities during the working hours. Thus, many experience that the boundaries between the spheres are less clear than before, which may have a

K. Grødal (✉)
Department of Public Health and Nursing, Faculty of Medicine and Health Sciences, Norwegian University of Science and Technology, PO Box 8905, 7491 Trondheim, Norway
e-mail: karoline.grodal@ntnu.no

M. Christensen et al. (eds.), *The Positive Side of Occupational Health Psychology*,
https://doi.org/10.1007/978-3-319-66781-2_6

further impact on how one experiences leisure time. Based on this, it would be beneficial to view the spheres as interacting rather than independent of one another (Geurts & Demerouti, 2003). We will see later in this chapter that the boundaries between work and family can be bound to challenges and negative effects, but also to possibilities and positive outcomes.

The literature of work and non-work contains numerous different concepts. For example, researchers define and delineate the non-work-related sphere in various ways. In this chapter, we concentrate on family life and, in accordance with Innstrand (2010), we examine the concept in a broader sense to capture life at home. Not everyone is a member of a family in its traditional form, and a broad understanding of the concept makes it more relevant for individuals in different family situations. The notion "home" is also often used with the same meaning. A great deal of the research conducted in this field has focused on the home and family life, and it will be made clear when we refer to research using different conceptualizations in this chapter. For example, some argue that a focus on family life gives a simplified picture of activities beyond the work sphere and therefore choose to use broader concepts such as life and non-work (e.g., Guest, 2002). We recognize that life consists of more than work and family, but still focus this chapter on the interaction between these spheres because they are likely to be the two most consuming spheres for most people in today's society. Leisure time will also be discussed as a relevant concept in the context of the theme. This is especially important since Norwegians increasingly value their spare time more. For example, it is more common than previously to start the weekend on Thursday or to work fewer hours on Friday (Statistics Norway, 2012).

6.1 What Is Work–Family Balance?

Early research on work–family balance mainly concerned negative aspects, and the balance was viewed as a lack of interference or conflict between the spheres. In recent research, positive aspects of the work–family balance, such as facilitation, have also caught attention and brought about a greater nuance of the concept. For example, this is reflected in the definition by Greenhaus, Collins, and Shaw (2003), who described work–family balance as "the extent to which an individual is equally engaged in—and equally satisfied with—his or her work role and family role" (p. 513). They further elaborate that important factors to consider are balance in time spent, psychological involvement, and satisfaction with the roles. Greater psychological involvement in working life than in the family is hence an indication of unbalance. The authors also differentiate between positive and negative balance, where equally high satisfaction in both roles represents a positive balance and equally low satisfaction in both roles represents a negative balance. A good work–family balance seems to be related to quality of life. Greenhaus et al. (2003) found that this especially applied to those who spend a relatively long time at work. The results showed that those spending more time on family than on work experienced a higher quality of life than those with a balanced timetable. The lowest quality of life was found among those who spent more time on work than family. An explanation could be that this group also experienced more work–family conflict and stress.

Today, it is common for researchers to take the effect and direction dimensions into account when dealing with work–family balance, and Frone (2003) suggested placing these in a taxonomy, as shown in Fig. 6.1. The spheres can affect each other in both directions, as displayed in the direction dimension. Work can influence family life (work-to-family) and family life can influence work (family-to-work), or the spheres can influence each other reciprocally. The effect dimension shows that the effect on balance can be either positive or negative, termed conflict or facilitation. Again, we find the concept of conflict that previously dominated the literature. The inclusion of facilitation shows that balance now is considered as more than just the absence of negative factors. Furthermore, we will take a closer look at the concepts of conflict and facilitation before summarizing the research on influence from work on family life, and the other way around.

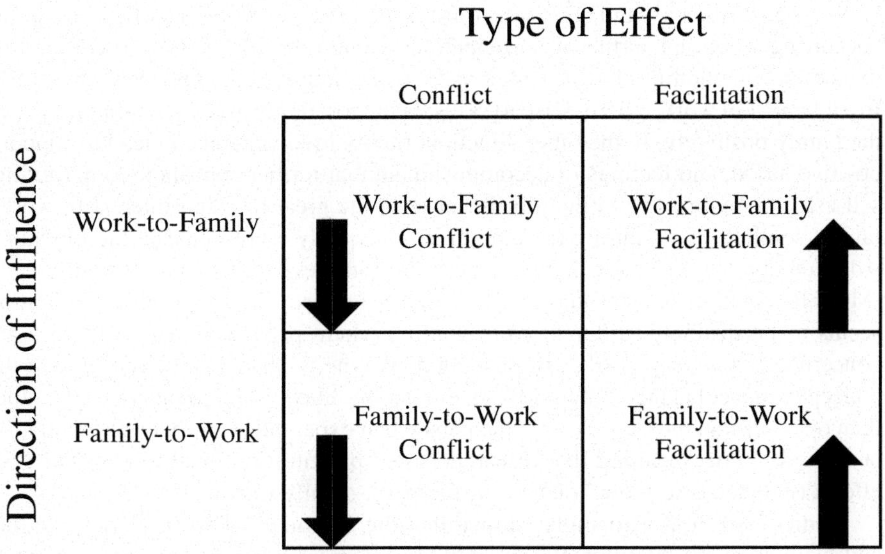

Fig. 6.1 Dimensions in work–family balance. *Source* Frone (2003, p. 146)

6.2 Conflict Versus Facilitation

Greenhaus and Beutell (1985) defined work–family conflict as “a form of interrole conflict in which the role pressures from the work and family domains are mutually incompatible in some respect” (p. 77). They mean that conflict occurs when participation in the family role becomes difficult because of the participation in the working role, and/or the reverse. They further suggested that important origins of conflict could be time involvement, strain, or behavioral demands connected to one of the roles. For example, work-to-family conflict may occur if a person’s time and energy spent at work compromises desired participation in family life. Conversely, a source of family-to-work conflict can be when difficulties in family life distract and challenge the work tasks. Greenhaus and Beutell’s (1985) definition has been, and still is, often used by researchers in the field of work and organizational psychology. Work–family conflict has been a popular research theme for decades, and studies have found many relationships between different precursors and outcomes. As we shall return to, the consequences have been shown to be unfavorable for the individual, especially in cases of conflict directed from work to family.

Wayne, Musisca, and Fleeson (2004) defined work–family facilitation as “occurring when, by virtue of participation in one role (e.g., work), one’s performance or functioning in the other role (e.g., family) is enhanced” (p. 110). Involvement in work can give rise to skills, behavior, or positive mood that influence the family positively. In the other direction, family involvement can lead to support, positive mood, and feelings of accomplishment that further contribute to mastering, self-esteem, and energy in the job. Because of the great research interest in work–family conflict, work–family facilitation is not equally represented in the literature. Additionally, the measures that have been used in these studies have varied greatly (McNall, Nicklin, & Masuda, 2010). Studies indicate, however, that facilitation seems to be distinctly different from conflict where precursors and outcomes are concerned (e.g., Grzywacz & Butler, 2005; Wayne et al., 2004). Even though the concepts impact balance differently, they are more likely orthogonal concepts rather than opposites on the same scale. One person can experience conflict and facilitation at the same time, meaning that researchers and practitioners need to consider both effects to understand the wholeness of the work–family interaction.

Studies have, unsurprisingly, shown that the outcomes of work–family conflict and facilitation are different. While conflict is connected to negative consequences, facilitation seems to be related to more favorable outcomes. Allen, Herst, Bruck, and Sutton (2000) reviewed several studies on outcomes of work-to-family conflict, and particularly large associations were found with burnout and work-related stress. Significant relationships were also found with factors such as lower job satisfaction, lower job commitment, turnover intentions and actual turnover, and lower satisfaction with life. A Norwegian study showed that conflict and facilitation were related, respectively, to an increased and reduced level of burnout at a later time (Innstrand, Langballe, Espnes, Falkum, & Aasland, 2008). The interaction from work to family was particularly important. In addition, the effect was reciprocal, indicating that the degree of burnout could be related to later conflict and facilitation. Later studies conducted on Norwegian church ministers and physicians showed that work–family interaction was more influential on later burnout than other factors such as the degree of autonomy and consistency between personal and work values (Innstrand, Langballe, & Falkum, 2011; Langballe, Innstrand, Aasland, & Falkum, 2011).

A study by van Steenbergen and Ellemers (2009) investigated the relationships of objective measures with health and found that work-to-family conflict was connected to high cholesterol and overweight. Similar findings were not found regarding family-to-work conflict. Facilitation was related to increased physical stamina, lower body mass index, and less likelihood of high cholesterol and overweight. Bakker, Shimazu, Demerouti, Shimada, and Kawakami (2013) conducted a study among Japanese parents of preschoolers where both parents were working. They based the study on principles from the job demands-resources model, which was described in Chap. 3, and found support that workaholism was correlated with job–family conflict, which further had an indirect negative effect on both a person's own and their partner's family satisfaction, also a year after. However, opposite results were found regarding work–family facilitation, which influenced the family satisfaction of both partners in a positive sense. Other studies have found that work–family facilitation is related to improved well-being (Allis & O'Driscoll, 2008), and less absenteeism and better performance at work (van Steenbergen & Ellemers, 2009).

It has previously been assumed that factors in one sphere lead to effects in the other sphere (e.g., Frone, Yardley, & Markel, 1997). According to this assumption, work-to-family conflict will, for example, influence satisfaction with the family life, but not with work. Later research has shown, however, that this is not necessarily the case, implying that the relationships are more complex. Kossek and Ozeki (1998) conducted a meta-analysis where they found that conflict, in both directions, had an impact on job satisfaction as well as satisfaction with life. Work-to-family conflict had the greatest effects on both outcomes. Wayne et al. (2004) found a pattern that indicated that conflict was associated with affective outcomes in the sphere the conflict originated from, and behavioral outcomes in the other. One example was that family-to-work conflict was related to lower satisfaction with family life and less effort in the job.

To understand how to promote a positive work–family balance, it is essential to know which factors are related to and create conflict and facilitation. Research on this will be presented in the following section. Innstrand, Langballe, and Falkum (2010) studied the prevalence of work–family conflict and facilitation in different occupational groups in Norway and found that bus drivers, teachers, and nurses experienced a great interaction between the spheres, while church ministers, lawyers, and people working in advertising did not. Innstrand, Langballe, Espnes, Aasland, and Falkum (2010b) examined how different family structures were related to work–family balance. Work–family conflict, in both directions, was experienced to a greater degree by parents of a nuclear family (two parents with children) than by single persons or childless couples. Single parents did not differ from the nuclear families when dealing with work-to-family conflict but reported greater family-to-work conflict. Corresponding results were found in a meta-analysis by Byron (2005), which showed that being single *with* children was connected to greater conflict than being single *without* children. Innstrand and colleagues (2010b) found a different pattern concerning facilitation. Childless couples reported higher levels of family-to-work facilitation than the nuclear families, while the family structure did not seem to impact work-to-family facilitation. It also seems that genders differ greatly when it comes to the interaction between work and family, and most studies have indicated that women experience both conflict and facilitation on a greater scale than men (e.g., Eby, Casper, Lockwood, Bordeaux, & Brinley, 2005; Innstrand et al., 2009; van Steenbergen, Ellemers, & Mooijaart, 2007).

Personal disposition also seems to influence the work–family balance. This has been shown, for example, in terms of characteristics regarding coping and effectiveness, factors that can contribute to a better interplay between the different spheres in life. Studies on the connection between the Big Five personality traits have shown that conscientiousness is associated with a smaller degree of work–family conflict and a greater degree of work-to-family facilitation (Michel & Clark, 2012; Wayne et al., 2004). In a meta-analysis, Byron (2005) found that a good coping style and the ability to manage time were related to a lower degree of work–family conflict in both directions. Primary self-evaluation, a concept incorporating self-esteem, self-efficacy, locus of control, and emotional stability, is another factor that seems to have an influence. People with good primary self-evaluation perceive that they have control over their own life, something that can give rise to more facilitation and less conflict (Michel & Clark, 2012). Positive and negative affect has also been shown to have a connection with facilitation and conflict (Michel & Clark, 2012). A Norwegian study showed that performance-based self-esteem was related to both conflict and facilitation, especially to work-to-family conflict (Innstrand, Langballe, Espnes, Aasland, & Falkum, 2010a). This result is worth noticing due to the tendency of the western society to appreciate the individual performance.

Research on the precursors of work–family conflict has to a great degree exceeded research on the precursors of work–family facilitation. The earlier mentioned meta-analysis by Byron (2005) comprised over 60 studies that had investigated precursors of conflict. Work-related factors, such as job stress, job involvement, workload, low support from colleagues and supervisors, and inflexible working hours, had a greater impact on work-to-family conflict than on family-to-work conflict. Job stress influenced employees with children more than those without children. Family-related factors like family stress, family conflict, age of the youngest child, and whether the partner was working had equally strong relationships with both directions of conflict, while the number of children and relationship status were of greatest importance for family-to-work conflict. The study also showed that the more time spent on family, housework, childcare, or other activities besides work, the greater the level of family-to-work conflict experienced. A more recent meta-analysis by Michel, Kotrba, Mitchelson, Clark, and Baltes (2011) showed similar results, and found a pattern in that most causes of conflict stemmed from the originating domain. For example, a great workload had a greater influence on work-to-family conflict than on family-to-work conflict.

As stated earlier, relatively little research has been conducted on work–family facilitation. In the direction from work to family, work engagement seems to contribute to positive affect, which further increases facilitation (Culbertson, Mills, & Fullagar, 2012). In the mentioned study, they found that the more a person had conversations with their partner about positive experiences in the workplace, the stronger was the connection between work engagement and work-to-family facilitation. A study by Grzywacz and Butler (2005) found that job resources, such as variation and autonomy, were important for work-to-family facilitation. Demands for social skills in the job benefited family life, probably because improving such skills makes one capable of mastering relations better, including at home. Personal growth was also an influential resource for work-to-family facilitation. One can imagine that growth and resources in the job combine with resources on the home front, creating a growth spiral as described in the theory of job demands-resources model and COR (see Chap. 3). It has been argued that conflict and facilitation can be explained based on the COR theory, since research indicates that stress and loss of resources lead to conflict, while a gain in resources increases facilitation (e.g., Innstrand, 2009; Wayne, Grzywacz, Carlson, & Kacmar, 2007).

6.3 Individual Strategies for Promoting Balance

Research results showing that women, compared to men, experience a higher level of both work–family conflict and facilitation were mentioned earlier in this chapter (Eby et al., 2005; Innstrand et al., 2009; van Steenbergen et al., 2007). A common explanation of why women experience more conflict has been that women work more than previously, while they are still left with most of the responsibilities in terms of home duties and childcare. However, since studies have shown that women also experience more facilitation, one can think of another explanation—namely, that women have more unclear boundaries between work and home so that the domains interact more, for better or worse. This could potentially be related to role identity connected to the different genders (Innstrand et al., 2009). A study on 115 female high-chool students who were married and had children was conducted at the end of the 1970s (Beutell & Greenhaus, 1983). Among these, approximately 70 percent experienced at least one form of work–family conflict. The researchers looked at different strategies for coping with this conflict, and it turned out that the most common was striving to meet expectations in all areas. However, attempts at being a supermom by being effective and making time for everything was the strategy with which fewest of the mothers succeeded. The two other coping mechanisms were more effective and were about either actively trying to influence the sources of the demands and thereby lower the expectations, or changing their own attitudes and their perceptions of others' expectations. These study results can possibly be explained by the idea that increased fulfillment of expectations may lead to increased demands. If the person concerned does not actively intervene and adjust this, one can eventually end up with a multitude of demands that are difficult to handle. Even though it has been over 30 years since this study was conducted, the results may still be relevant today in light of the public debate on the emerging "perfection generation". Many are concerned about the (un)healthiness of the increasing trend of striving to achieve and succeed in several areas simultaneously. This can be a perception that, to be good enough, performing well in the job, being a perfect parent, looking good, exercising, eating healthily, and having several hobbies are expectations that should all be met.

Some argue for differentiation between integration and segmentation as individual strategies for dealing mentally with boundaries between work and leisure time (e.g., Nippert-Eng, 1996). Integration means viewing the spheres as intertwined, while segmentation is to draw clear boundaries between them. As mentioned in the introduction of this chapter, information and communication technology is becoming increasingly important for how we live our lives, both at work and home (Bouwman et al., 2005). A good example of this is the smartphone, which gives the possibility of being connected to the Internet and different applications, meant for both work and private use, almost anytime and in any place. It gives great possibilities for a flexible and practical daily life, but at the same time, it contributes to making the distinction between work and leisure time unclear. This promotes integration and may eventually lead to challenges. More people sacrifice their leisure time in the evenings and at weekends and may find it difficult to disconnect completely from work and to have real time off. Several Icelandic academics reported on this in a qualitative study, and the results showed that women experienced a greater challenge with this than men (Heijstra & Rafnsdottir, 2010). Employees are thus experiencing new ways of working, and these can make it difficult to extract themselves from work outside working hours. For many, it can be useful to make efforts to segregate the spheres to enable concentration at home or other non-work domains and to ensure good recovery and stamina at work. One example is to shut down the work phone and the computer after finishing the workday and to have a separate phone for personal use (e.g., Hill, Miller, Weiner, & Colihan, 1998). Others may turn off notifications and have specific times during the day for checking e-mails, which can, however, be difficult for those who experience real or self-inflicted demands to be available at any hour.

Family life does not necessarily overlap with leisure time (Geurts & Demerouti, 2003). Firstly, one can have leisure time activities unrelated to family. Secondly, for some employees, family obligations may feel more like work than leisure time. The content of leisure time varies between individuals, and some employees may experience the workplace as a sanctuary from a stressful situation at home. This emphasizes that an individual's overall life situation is important to consider for coping and achieving a good balance. For some, work–family balance in time implies balance in satisfaction. Others may have coping strategies where they periodically sacrifice the time balance with the intention that satisfaction will be improved over time. One employee may, for example, work more in busy periods to attain control in the job and therefore be able to fully let go of the work tasks and be focused on other spheres when he or she is actually off duty. Karasek (1976) pointed at a tendency that employees generally carry social patterns with them from the job into leisure time. Thus, it seems that people generally do not compensate for an active working life by being passive outside work, but on the contrary, they are also active in their leisure time.

6.4 Organizational Initiatives for Promoting Balance

In Norway, there is a relatively generous public family policy compared to other countries (Gulbrandsen & Jensen, 2009). There are several welfare programs that intend to promote a family-friendly working life. Examples are flexible programs for parental benefits and leave: fathers are entitled to two weeks of paid leave in conjunction with births or adoption, and each parent is entitled to ten days' leave with their sick children. Several private companies also gradually take on social responsibilities and offer their employees welfare programs on top of the public system (Gulbrandsen & Jensen, 2009). This can include flexible working hours, possibilities for home office, compressed working weeks, job sharing (sharing a full-time job with someone else), kindergarten at the workplace, and/or an extra week's vacation with pay. It is common to view these family-friendly arrangements as serving organizational objectives such as effectiveness (Beauregard & Henry, 2009). The efforts can contribute to better-performing employees, but also to a positive reputation among customers, and can create a more attractive workplace in the competition for recruiting the best employees. As we have seen earlier in this chapter, employees with children are the most vulnerable to conflict between work and family, and it may therefore be extra important to have measures aimed at benefiting this group.

Research on the effects of family-friendly programs has often applied measurements examining whether employees have access to the programs, and not whether the employees actually use them (Beauregard & Henry, 2009). A study in a big American cooperation showed that even though the employees were offered beneficial programs, few actually made use of them (Hochschild, 1997). Bø (2006) conducted a study in three Norwegian companies where company leaders, employees with children, and the employees in the children's kindergarten were interviewed. The results showed there was a difference in the companies' communicated ideals and actual practice when dealing with family-friendly programs. Some mothers experienced negative reactions from leaders if they made use of their right to stay at home with their sick child. It was also emphasized that programs aimed at, for example enhancing employees' flexibility can become a challenge when they are present but are too limited to solve family challenges. If a kindergarten opens at 7:30a.m, it does not help the employee to be offered the opportunity to start work at 7:30a.m if the purpose is to give them the possibility of following their child. The employee will be stuck with the same challenge, and when one has finally been offered a personal arrangement, it can give a sense that one should be grateful and not ask the employer for more. The same study also showed that programs were perceived to be more favorable toward men than women, despite today's official norm that employees of both genders shall have equal opportunities in their working life. This is interesting since women also experience more conflict than men. Is it the case that women and men actually need to be treated differently in efforts to achieve a good balance? Do women, due to home responsibilities, generally, have a greater total load that makes it more challenging to combine family with a career? What can possibly be done about this? These are questions that can inspire future research on how a good work–family balance can be promoted for both genders.

When an organization implements measures to increase work–family balance, it is important to be aware of the effects that are aimed at. One single measure can, for example, have a positive influence on work-to-family conflict, but worsen the conflict from family toward work. A study by Golden, Veiga, and Simsek (2006) showed that the more the employees worked from home, the more family-to-work conflict and less work-to-family conflict they experienced. Autonomy and flexible working hours seemed to moderate the effect on family-to-work conflict, but home office still appeared to be a solution that was more family-friendly than it was convenient regarding work. Nevertheless, organizations probably have much to gain from facilitating employees' well-being, including outside work. If this contributes to a healthy workforce, the organizational outcome may be less absenteeism and a well-performing staff—a win-win situation for the organization, the employees, and their families.

6.5 Conclusion

This chapter's focus has been on positive and negative aspects of work–family balance. The "new work life" seems to contribute to increasingly unclear boundaries between work and leisure, which can represent challenges—for example, in order to be fully present in roles related to both work and family. A better understanding of how employees can cope with this may contribute to a more valuable leisure time, something that can also benefit work. For some, this can make a difference in enabling them to actually feel how it is to have a break. The topic is important and relevant in these times and will probably continue to be so. Work life and society are continually changing, and this requires updated knowledge about appropriate efforts and solutions. As discussed, this is important for the employees and their families' health and well-being, but it is also reasonable to assume that the organization and society as a whole can gain from employees who experience a good balance between work and family life. Healthy employees are essential for productivity and presence at work and can benefit society in the form of fewer expenses related to, for example, poor health, sickness absenteeism, and disability.

There is an evident lack of research on the positive aspects of work–family interaction, especially concerning what enhances facilitation. Studies looking at these aspects will be of great importance for gaining knowledge about how to facilitate employees in different life situations coping with everyday life, including demands from work, family life, and leisure time, in the best possible manner. Simultaneously, it will be important for both researchers and politicians to differentiate between and be aware of the dimensions of effect and direction when dealing with work–family balance (cf. Frone, 2003). This can help with obtaining more precise knowledge on the topic and succeeding with the intended aims of interventions. Research on the effects of interventions and measures and not only on whether employees are offered them will also be beneficial. Finally, it has been highlighted that few studies have investigated how family-related variables affect behavior at work (Eby et al., 2005). Instead of focusing only on the things that have to do with work, an understanding of the employees' holistic life situation can be a key for greater success in obtaining organizational goals such as productivity, lower sickness absenteeism, and a good and health-promoting working environment.

References

Allen, T. D., Herst, D. E. L., Bruck, C. S., & Sutton, M. (2000). Consequences associated with work-to-family conflict: A review and agenda for future research. *Journal of Occupational Health Psychology, 5,* 278–308. doi:10.1037//1076-8998.5.2.2.

Allis, P., & O'Driscoll, M. (2008). Positive effects of nonwork-to-work facilitation on well-being in work, family and personal domains. *Journal of Managerial Psychology, 23,* 273–291. doi:10.1108/02683940810861383.

Bakker, A. B., Shimazu, A., Demerouti, E., Shimada, K., & Kawakami, N. (2013). Work engagement versus workaholism: A test of the spillover-crossover model. *Journal of Managerial Psychology, 29,* 63–80. doi:10.1108/JMP-05-2013-0148.

Beauregard, T. A., & Henry, L. C. (2009). Making the link between work-life balance practices and organizational performance. *Human Resource Management Review, 19*(1), 9–22. doi:10.1016/j.hrmr.2008.09.001.

Beutell, N. J., & Greenhaus, J. H. (1983). Integration of home and nonhome roles: Women's conflict and coping behavior. *Journal of Applied Psychology, 68*(1), 43–48.

Boswell, W. R., & Olson-Buchanan, J. B. (2007). The use of communication technologies after hours: The role of work attitudes and work-life conflict. *Journal of Management, 33,* 592–610. doi:10.1177/0149206307302552.

Bouwman, H., van den Hooff, B., van de Wijngaert, L., & van Dijk, J. (2005). *Information & communication technology in organizations.* London: Sage.

Byron, K. (2005). A meta-analytic review of work-family conflict and its antecedents. *Journal of Vocational Behavior, 67,* 169–198. doi:10.1016/j.jvb.2004.08.009.

Bø, I. (2006). Working life and family life: Ambiguous communication at work. *Community, Work & Family, 9*(2), 123–141. doi:10.1080/13668800600586951.

Culbertson, S. S., Mills, M. J., & Fullagar, C. J. (2012). Work engagement and work-family facilitation: Making homes happier through positive affective spillover. *Human Relations, 65,* 1155–1177. doi:10.1177/0018726712440294.

Eby, L. T., Casper, W. J., Lockwood, A., Bordeaux, C., & Brinley, A. (2005). Work and family research in IO/OB: Content analysis and review of the literature (1980–2002). *Journal of Vocational Behavior, 66,* 124–197. doi:10.1016/j.jvb.2003.11.003.

Frone, M. R. (2003). Work-family balance. In J. C. Quick & L. E. Tetrick (Eds.), *Handbook of occupational health psychology* (pp. 143–162). Washington, DC: American Psychological Association.

Frone, M. R., Yardley, J. K., & Markel, K. S. (1997). Developing and testing an integrative model of the work-family interface. *Journal of Vocational Behavior, 50,* 145–167. doi:10.1006/jvbe.1996.1577.

Geurts, S. A. E., & Demerouti, E. (2003). Work/non-work interface: A review of theories and findings. In M. J. Schabracq, J. A. M. Winnubst, & C. L. Cooper (Eds.), *The handbook of work and health psychology* (pp. 279–312). Chichester, UK: Wiley.

Golden, T. D., Veiga, J. F., & Simsek, Z. (2006). Telecommuting's differential impact on work-family-conflict: Is there no place like home? *Journal of Applied Psychology, 91,* 1340–1350. doi:10.1037/0021-9010.91.6.1340.

Greenhaus, J. H., & Beutell, N. J. (1985). Sources of conflict between work and family roles. *Academy of Management Review, 10*(1), 76–88.

Greenhaus, J. H., Collins, K. M., & Shaw, J. D. (2003). The relation between work-family balance and quality of life. *Journal of Vocational Behavior, 63*(3), 510–531. doi:10.1016/S0001-8791(02)00042-8.

Grzywacz, J. G., & Butler, A. B. (2005). The impact of job characteristics on work-to-family facilitation: Testing a theory and distinguishing a construct. *Journal of Occupational Health Psychology, 10,* 97–109. doi:10.1037/1076-8998.10.2.97.

Guest, D. E. (2002). Perspectives on the study of work-life balance. *Social Science Information, 41*(2), 255–279. doi:10.1177/0539018402041002005.

Gulbrandsen, T., & Jensen, R. S. (2009). *Velferdstiltak i private bedrifter: En litteraturgjennomgang [Welfare measures in private sector organizations: A literature review].* Oslo: Institute for Social Research.

Halpern, D. F. (2005). Psychology at the intersection of work and family: Recommendations for employers, working families, and policymakers. *American Psychologist, 60*(5), 397–409.

Heijstra, T. M., & Rafnsdottir, G. L. (2010). The internet and academics' workload and work-family-balance. *Internet and Higher Education, 13,* 158–163. doi:10.1016/j.iheduc.2010.03.004.

Hill, E. J., Miller, B. C., Weiner, S. P., & Colihan, J. (1998). Influences of the virtual office on aspects of work and work/life balance. *Personnel Psychology, 51,* 667–683. doi:10.1111/j.1744-6570.1998.tb00256.x.

Hochschild, A. R. (1997). *Time bind. When work becomes home & home becomes work.* New York: Metropolitan Books.

Innstrand, S. T. (2009). *Work/home conflict and facilitation: COR(e) relations (Unpublished doctoral dissertation).* Trondheim: Norwegian University of Science and Technology.

Innstrand, S. T. (2010). Betydningen av en god balanse mellom arbeid og familieliv [The importance of a good balance between work and family life]. *Søkelys på arbeidslivet, 26*(1–2), 31–41.

Innstrand, S. T., Langballe, E. M., Espnes, G. A., Falkum, E., & Aasland, O. G. (2008). Positive and negative work-family interaction and burnout: A longitudinal study of reciprocal relations. *Work & Stress, 22*(1), 1–15. doi:10.1080/02678370801975842.

Innstrand, S. T., Langballe, E. M., Espnes, G. A., Aasland, O. G., & Falkum, E. (2010a). Personal vulnerability and work-home interaction: The effect of job performance-based self-esteem on work/home conflict and facilitation. *Scandinavian Journal of Psychology, 51,* 480–487. doi:10.1111/j.1467-9450.2010.00816.x.

Innstrand, S. T., Langballe, E. M., Espnes, G. A., Aasland, O. G., & Falkum, E. (2010b). Work-home conflict and facilitation across four different family structures in Norway. *Community, Work & Family, 13*(2), 231–249. doi:10.1080/13668800903314341.

Innstrand, S. T., Langballe, E. M., & Falkum, E. (2010c). Exploring occupational differences in work-family interaction: Who is at risk? *International Journal of Stress Management, 17*(1), 38–55. doi:10.1037/a0018565.

Innstrand, S. T., Langballe, E. M., & Falkum, E. (2011). The longitudinal effects of individual vulnerability, organisational factors, and work-home interaction on burnout among male church ministers in Norway. *Mental Health, Religion & Culture, 14*(3), 241–257. doi:10.1080/13674670903470621.

Innstrand, S. T., Langballe, E. M., Falkum, E., Espnes, G. A., & Aasland, O. G. (2009). Gender-specific perceptions of four dimensions of the work/family interaction. *Journal of Career Assessment, 17,* 402–416. doi:10.1177/1069072709334238.

Karasek, R. A. (1976). *The impact of the work environment on life outside the job (Unublished doctoral dissertation).* Cambridge: Massachusetts Institute of Technology.

Kossek, E. E., & Ozeki, C. (1998). Work-family conflict, policies, and the job-life satisfaction relationship: A review and directions for organizational behavior-human resources research. *Journal of Applied Psychology, 83,* 139–149. doi:10.1037//0021-9010.83.2.139.

Langballe, E. M., Innstrand, S. T., Aasland, O. G., & Falkum, E. (2011). The predictive value of individual factors, work-related factors, and work-home interaction on burnout in female and male physicians: A longitudinal study. *Stress & Health, 27*(1), 73–87. doi:10.1002/smi.1321.

McNall, L. A., Nicklin, J. M., & Masuda, A. D. (2010). A meta-analytic review of the consequences associated with work-family enrichment. *Journal of Business and Psychology, 25,* 381–396. doi:10.1007/s10869-009-9141-1.

Michel, J. S., & Clark, M. A. (2012). Investigating the relative importance of individual differences on the work-family interface and the moderating role of boundary preference for segmentation. *Stress and Health, 29,* 324–336. http://dx.doi.org/10.1002.smi.2474.

Michel, J. S., Kotrba, L. M., Mitchelson, J. K., Clark, M. A., & Baltes, B. B. (2011). Antecedents of work-family conflict: A meta-analytic review. *Journal of Organizational Behavior, 32,* 689–725. doi:10.1002/job.695.

Nippert-Eng, C. (1996). Calendars and keys: The classification of "home" and "work". *Sociological Forum, 11*(3), 563–582.

Roos, P. A., Trigg, M. K., & Hartman, M. S. (2006). Changing families/changing communities. *Community, Work & Family, 9,* 197–224. doi:10.1080/13668800600586985.

Statistics Norway. (2012). *Time use survey, 2010.* Retrieved from http://www.ssb.no/en/kultur-og-fritid/statistikker/tidsbruk.

Statistics Norway. (2013). *Families and households, 1 January 2013.* Retrieved from http://www.ssb.no/en/befolkning/statistikker/familie.

Statistics Norway. (2014). *This is Norway 2014. What the figures say.* Retrieved from http://ssb.no/en/befolkning/artikler-og-publikasjoner/this-is-norway-2014.

van Steenbergen, E. F., & Ellemers, N. (2009). Is managing the work-family interface worthwhile? Benefits for employee health and performance. *Journal of Organizational Behavior, 30,* 617–642. doi:10.1002/job.569.

van Steenbergen, E. F., Ellemers, N., & Mooijaart, A. (2007). How work and family can facilitate each other: Distinct types of work-family facilitation and outcomes for women and men. *Journal of Occupational Health Psychology, 12,* 279–300. doi:10.1037/1076-8998.12.3.279.

Wayne, J. H., Grzywacz, J. G., Carlson, D. S., & Kacmar, K. M. (2007). Work-family facilitation: A theoretical explanation and model of primary antecedents and consequences. *Human Resource Management, 17,* 63–76. doi:10.1016/j.hrmr.2007.01.002.

Wayne, J. H., Musisca, N., & Fleeson, W. (2004). Considering the role of personality in the work-family experience: Relationships of the big five to work-family conflict and facilitation. *Journal of Vocational Behavior, 64,* 108–130. doi:10.1016/S0001-8791(03)00035-6.

Chapter 7
Work Engagement and Job Crafting

Marit Christensen

In order for employees to feel good, have good health and be productive, a constructive psychosocial working environment is important. (Bakker & Demerouti, 2014; Christensen, 2008; Christensen, Aronsson, Clausen, Hakanen, & Vivoll Straume, 2012). Good leadership, a well-designed job, and working conditions that promotes work engagement and good performance, is important in achieving this. Organizations usually initiate top-down governed interventions to increase motivation, to decrease sickness absenteeism, and to increase performance. However, they do not always get the expected results (Aust, Rugulies, Finken, & Jensen, 2010). Many organizations also find that the bottom-up-processes of engagement and performance, driven by the employees themselves, combined with equivalent processes driven by the management, to be more useful. One kind of bottom-up processes is job crafting, where the employees themselves form the job in such a way that it is perceived as more engaging. Leadership can stimulate job crafting by promoting the right work conditions for their employees. Job crafting can be understood as a type of proactive behavior where the employees themselves takes the initiative to change the level of demands and resources to make their job more meaningful, engaging, and satisfying. In this chapter, we will first explain what work engagement is, as well as the theoretical background. Second, we will examine the concept of job crafting. Finally, we will give some tips on how one can work with job crafting in practice.

M. Christensen (✉)
Department of Psychology, Norwegian University of Science and Technology (NTNU), Trondheim, Norway
e-mail: marit.christensen@ntnu.no

M. Christensen et al. (eds.), *The Positive Side of Occupational Health Psychology*,
https://doi.org/10.1007/978-3-319-66781-2_7

7.1 Work Engagement

Today, the leading definition of work engagement is "… a positive, fulfilling, work related state of mind that is characterized by vigor, dedication, and absorption" (Schaufeli & Bakker, 2010; Schaufeli, Salanova, Gonzales-Roma, & Bakker, 2002, p. 74). This definition implies that it is not a specific short-term experience of an emotion, but a lasting affective-cognitive condition. These three dimensions represent different aspects of the conception of work engagement. Vigor deals with having energy and the will to put effort into one's work, and having endurance when dealing with challenges. Dedication concerns a strong involvement and identification with the job, furthermore it deals with being inspired and proud of one's work, as well as thinking of the work as important. The ability to be absorbed in one's work, concerns an experience of deep concentration and a sense of time passing quickly, and having trouble with leaving the work. Today, the research community is in relative agreement with the notion that work engagement includes a behavioral component (vigor), an emotional component (dedication), and a cognitive component (ability to be absorbed in the job). The most commonly used measurement when measuring work engagement is the Utrecht Work Engagement Scale (UWES: Schaufeli & Bakker, 2003). The instrument consists of 17 questions and has also a shorter version consisting of nine questions.

Research has shown that both job related resources and personal resources influence engagement at work. Even though different professions and jobs differ in determining which demands and resources are the most significant, we find some common features (Halbesleben, 2010; Mauno, Kinnunen, Mäkikangas, & Feldt, 2010; Crawford, LePine, & Rich, 2010; Christian, Garza, & Slaughter, 2011). Meta-analyses show that the experience of autonomy and freedom in the workplace has a great impact on the work engagement. Furthermore, we find factors such as perceived influence and participation in decision making as important predictors for work engagement. Social components such as social support from management and co-workers, both emotionally, as well as professionally, are important for work engagement. Additionally, research shows that teamwork, feedback, and recognition create engagement. Constructive feedback on the work done, and recognition of good work, has an important impact. Another important factor for work engagement deals with the opportunity for learning and personal development, that again gives rise to personal growth. Factors, such as task variety, being given and holding responsibility, perceived justice and value adjustments between personal values and organisational values, are also promoters of work engagement. A transformational leadership usually considers the context of work engagement. This form of leadership seems to be especially favorable in promoting work engagement amongst the employees. The theory of transformational leadership emphasizes values and vision. The leader is a role model, a motivator through inspiration, a visionary, and a challenger of new thoughts who sees the unique needs of every single worker (Hetland, 2008). Last but not least, research shows that challenges are also important in creating engagement. Employees need demands and challenges in order to be engaged, as well as their need for promoting some of the mentioned job recourses.

Why is work engagement that important? Research has gathered considerable empirical evidence that supports several favorable outcomes of work engagement. Engaged workers have been found to have a lower level of depression (Hakanen & Schaufeli, 2012), a more positive experience of mental/physical health (Schaufeli, Taris, & van Rhenen, 2008), better sleep quality (Kubota et al., 2011), and better work ability (Airila et al., 2014). There is support for the view that work engagement has a positive association with a variety of organizational outcomes. One study showed that work engagement amongst hotel and restaurant employees in Spain gave better costumer evaluations of service (Salanova, Agut, & Peiró, 2005). Another study conducted on leaders and workers considered the work performance as better in employees who were showing more engagement (Halbesleben & Wheeler, 2008). Moreover, studies have shown engaged workers to be more innovative (Hakanen, Perhoniemi, & Toppinnen-Tanner, 2008), and also that work

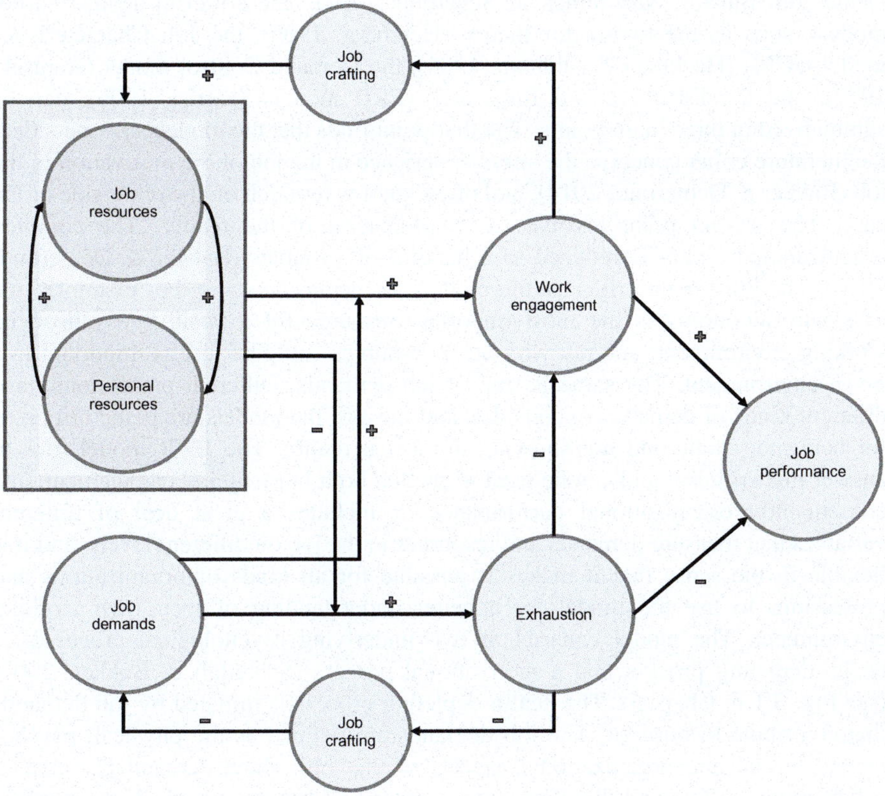

Fig. 7.1 Job Demand-Resources theory including job crafting. *Source* Bakker and Demerouti (2014)

engagement has an effect on the financial returns of an organization (Xanthopoulou, Bakker, Demerouti, & Schaufeli, 2009). In a longitudinal study, it was found that work engagement had a relationship with sickness frequency (Schaufeli, Bakker, & Van Rhenen, 2009). This means that engaged workers have a lower sickness frequency than others. One should also note that work engagement could have negative aspects, since excessive engagement can affect the work-family balance negatively. Most of the jobs that create work engagement and the possibility for personal growth are complex and demanding jobs that can also lead to stress, burnout, and overload. The most commonly used theoretical model for surveying work engagement empirically is the Job-Demand-Resources theory (JD-R) (Fig. 7.1).

7.2 Job-Demand-Resources Theory (JD-R)

As a result of the criticism of the earlier stress and motivation models, the Job-Demand-Resources model (JD-R) was developed (Bakker & Demerouti, 2007; Demerouti, Bakker, Nachreiner, & Schaufeli, 2001). The criticism against earlier models, such as the two-factor theory (Herzberg, 1966), the Job Characteristics model (JCM) (Hackman & Oldham, 1980), the Demand-Control model (Karasek, 1979), and the Effort-Reward Imbalance (ERI) model (Siegrist, 1996), can be summarized in three main points. The first point finds that the models are one-sided; the literature either concerns the stress experience in the job or the motivation in the job (Bakker & Demerouti, 2014), and often ignores research on the other side of the case. The second point involves a simplification of the reality. The complex organizational reality is reduced to a handful of variables (Bakker & Demerouti, 2014). The third point criticizes the models for being too static. For example, one asks why autonomy is the most important resource for a positive psychosocial working environment, and not why other resources could be just as important in a given environment. The same is true of job demands. Different professions have different kinds of demands to consider, and most of the models are being criticized for being too static and not showing the actual reality. The JD-R model tries to answer this criticism and can be used to predict both burnout and engagement, and consequently organizational performance. It includes a great deal of different variables that measure demands and resources in the job on different levels, and it is flexible in the sense that it makes it possible for all kinds of organizations and professions to use the model for a better understanding of their own working environment. The model consists of two underlying psychological processes, a health-depleting process and a motivational process (Schaufeli & Bakker, 2004) (see Fig. 3.1 in Chap. 3). The health-depleting process is initiated by job demands that contribute to burnout. The job demands are defined as the physical, psychological, social and organizational aspects of the job, which demand physical or psychological effort and therefore are associated with some physical and psychological costs. This could, for example, be a heavy workload, time pressure, bad physical work environment, work-family conflict, and emotionally demanding interactions with students, patients, clients, or customers. Not all job demands are necessarily negative. Some demands can be experienced as obstructive such as emotional demands and work-family conflicts, whilst other demands can be challenging such as time pressure and workload (Crawford et al., 2010). The motivational process is initiated by job resources that through this process contribute to work engagement. Job resources are defined by the physical, psychological, social, or organizational aspects of the job that (1) are functional when achieving goals, (2) reduce job demands and the associated psychological and physical cost, and (3) stimulate personal growth, learning and development. There is also an interaction between the two parallel processes. Job resources work as a buffer for the influence of job demands on burnout. Job resources have a special influence on motivation and job engagement when the job demands are high (Bakker & Demerouti, 2007; Hakanen, Bakker, & Demerouti, 2005; Bakker, Hakanen, Demerouti, & Xanthopoulou, 2007). Flexibility is also an important keyword in the JD-R-model. One postulates that both motivational processes and the health-depleting processes are independent of the demands and recourses we find in the model. This means that we can use the model across different organizations and professions (Bakker & Demerouti, 2007). Good job design and good leadership facilitate the motivation of employees and reduce the stress experienced. However, these working conditions are not always present. It could also happen that employees see some potentially positive changes that could be made to their situation. In these situations, one can initiate measures where the employee can partake actively and change the job design through job crafting by making different changes to different tasks, contents, and relations, in this way creating more meaning in their work.

7.3 Job Crafting

Job crafting is a relatively new research field, gradually gathering more empirical evidence. The basic principles of job crafting can exist side by side with a top-down process in creating greater engagement and performance in responding to complexity in work and in meeting the needs of today's employees (Demerouti, 2014). This indicates something other than employees' participation in redesigning the job, and something more than proactive behavior. Tims, Bakker, and Derks (2012) found that while proactivity is directed towards exceeding the accomplishments in the job, job crafting is rather directed towards enhancing one's own motivation and person-environment fit. There is an important difference between job design (top-down) and job crafting (bottom-up). In job design, the leaders have the responsibility for achieving the optimal adjustments between the employees and the job; whilst in job crafting the workers themselves initiate the changes. The focus of job crafting is that the employees adapt and change work tasks to optimize their workday, while the focus of job design deals with satisfying the needs of employees to create better efforts (Lyons, 2008; Wrzesniewski & Dutton, 2001). Job crafting can often seem like an inner action, not visible to the other co-workers or to the leaders, and that has to be learned (Wrzesniewski & Dutton, 2001). Job crafting is also a part of a process that develops through time (Berg, Dutton, & Wrzesniewski, 2013).

The premise for job crafting has to be a possibility of adjustment and making changes in the psychosocial work environment. The employees partake actively in the workplace by forming their own job tasks and social relations, and thereby the idea is that they are partaking in forming their psychosocial work environment (Tims et al., 2012). It should be noted that there are different degrees of freedom in different jobs concerning job crafting, and not everyone feels the need for job crafting, but, if one is dissatisfied with the job design, then job crafting can be a good alternative. By using job crafting, the workers put in a greater effort because of the possible positive outcome, not necessarily because they wish to work more, but because job crafting can increase productivity. The need for job control, positive self-image, and social contact with other people, motivates job crafting. In understanding and explaining why people craft in their job, it can be useful to take a look at the literature on motivation, goal setting theory, and expectation theory (see for example Hobfoll, 2001; Locke & Lathan, 1990; Vroom, 1964). Employees have a conception of what they wish for their job in the future, and thereafter develop different strategies for achieving this. Job crafting is driven by the motivation for achieving these goals.

7.4 Defining Job Crafting

Job crafting has been operationalized and measured with on the basis of two different approaches. The first approach was from of Wrzesniewski and Dutton (2001) who created the term. Wrzesniewski and Dutton (2001), define job crafting as physical and cognitive changes the employees make in relation to their tasks and relationships at work, as well as in relation to potential limitations of the job. Changes in physical limitations concern form, scope, and the number of work tasks, while changes in cognitive limitations concern changes in how one perceives the job. The relational limitations concern to what degree one cooperates with others, and with whom one cooperates. By changing these elements, one changes the design of the job and the social working environment (Demerouti, 2014). The second approach is based on the Job Demand-Resources model (JD-R) and involves the balance for employees between job demands and job resources with their own abilities and needs (Tims & Bakker, 2010). More specifically, one can say that job crafting is a more proactive behavior that consists of three dimensions: seeking resources, challenges, and the reduction of demands (Petrou, Demerouti, Peeters, Schaufeli, & Hetland, 2012). Seeking job resources can be seen as a form of mastery of the job demands one experiences, or as a functional completion of the tasks and goal achievement. Hobfoll (2001) supports this view with his theory of Conservation of Resources (COR) where he postulates that our motivation targets obtaining, maintaining and protecting the resources we value. Seeking new challenges can consist of finding new challenging work tasks, as well as taking more responsibility for maintaining motivation and avoiding boredom. Job crafting does not just consist of finding strategies for discovering favorable characteristics in the job. The third dimension is concerned with behavior directed towards reducing or minimizing the emotional, mental and physically demanding aspects of the job (Demerouti, 2014). The two approaches to job crafting are overlapping. Wrzesniewski and Dutton (2001) suggested that job crafting concerns changing one's job in order to discover a greater meaning, whilst the JD-R perspective focuses on the job characteristics that influence motivation and health amongst the employees (Tims & Bakker, 2010). Both models focus on the employee making decisions about changing their job by tackling problems better and by finding solutions. However, still the models are a bit different regarding their specific targets of job crafting.

Since job crafting is a relatively new area, much of the research been qualitative, but we also find quantitative measurements. Wrzesniewski and Dutton (2001) developed the first measurement of task crafting, relational crafting, and cognitive crafting. In 2012, Tims and colleagues developed an instrument that included four different dimensions: (1) increase in social job resources, (2) increase in structural job resources, (3) increase in challenging demands, and (4) decrease in hindering job demands.

7.5 Which Factors Create Job Crafting and What Are the Consequences?

Demanding jobs seem to stimulate proactive behavior, because task complexity is a predictor for job crafting (Ghitulescu, 2007). When employees experience a great deal of work related pressure and autonomy, there is a greater chance of job crafting. This was supported by Karasek's (1979) theory on active workers. Those with active jobs perform more resource seeking and are less oriented towards reducing demands in the Job-Demand-Control model.

Job crafting seems to be a healthy way to tackle changes in working life. Kira, Balkin, & San (2012) found that workers job craft their workplace relationships, for example, when asking for support from leaders, and job craft job tasks, for example when priorities are needed for tackling new job situations. Individual characteristics also had an impact on which level of job crafting is used. Employees with a proactive personality had a higher probability of operating with job crafting, meaning they strengthened their structural and social job resources, as well as increasing the job challenges (Bakker, Tims, & Derks, 2012). These elements should be viewed together, since the person-environment fit is essential when explaining job crafting behavior. Demerouti (2014) concluded that job crafting presents itself in demanding, resourceful work environments in constant change, consistent with workers who are proactive and motivated by growth, and/or who experience a mismatch between their motivational style and characteristics outside the work environment.

An important question in this context regards which level of job crafting is positive and effective. Wrzesniewski and Dutton (2001), when launching the concept, suggested that employees who crafted their job were more satisfied, because it represented an opportunity for promoting the employees' experiences of meaning of their work. Research has found a positive correlation between job crafting and organizational commitment (Ghitulescu, 2007), psychological well-being (Berg, Grant, & Johnson, 2010), work performance (Leana, Appelbaum, & Shevchuk, 2009; Tims et al., 2012), and the experience of greater meaning (Wrzesniewski, LoBuglio, Dutton, & Berg, 2013). (Figure 3.1 in Chap. 3 shows how job crafting is included in the JD-R model).

7.6 Practical Examples of Job Crafting

On a practical level, which methods can one use to operate effectively with job crafting? An important condition should concern the possibility of adapting and changing components in the psychosocial work environment. Awareness of what is meant by the notion of job crafting could be a start. One can arrange workshops on explaining the phenomenon and discussing the background of the JD-R model. Here, one can emphasize the win-win situation for both the management and the employees through promoting the engagement, motivation, health, and job performance of co-workers. When stimulating job crafting, it is essential to work on the climate, norms, and leaders' attitude towards this kind of initiative and to monitor the interventions.

One of the most important dimensions in job crafting involves increasing the resources. This can be dealt with by, for example, seeking feedback from one's leaders and co-workers and possibly contributing to developing areas for a natural flow of feedback and recognition. An example which emphasizes this could be to hold meetings to mark happy events and successes, as well as to give recognition to co-workers who have achieved different goals. Of course, this should be performed in a thoughtful and just manner. Furthermore, a job crafting initiative could actively seek support from one's leaders and co-workers. If one is clear on one's needs and gives back by positive collaboration, this could create engagement. Investing in one's relationships and spending time on this, is important in order to achieve the resources that can contribute to feedback, support, and acknowledgement. An approach is to use the freedom and autonomy one possesses in the job when cooperating with colleagues to give rise to an energetic and positive performance, as well as working more frequently with likeable tasks, again giving the experience of meaning.

Another important dimension of job crafting deals with increasing one's challenges. Here one can use one's talents more and follow one's interest when it seems possible. Questions of whether there is an unused potential in the job where one can continue to develop and to take advantage of abilities, are important. Professional development is often perceived as a predictor of work engagement (Bakker & Demerouti, 2008), and therefore it is important to use the learning opportunities one has in the job. It is also possible to hire external resources, such as a coach or a mentor who could contribute to new challenges and motivational factors.

The last dimension deals with simplification and reduction of demands. Here one can consider different initiations when working more effectively and avoiding procrastination through better planning and timing. Another important element, at which most of us can succeed, is delegating work to others, using the resources at hand and thereby letting go of perfectionism. Developing clear agreements with clear goals can be beneficial. In periods with deadlines and time pressure, searching for quiet places where one can work without being disturbed could be a smart solution.

7.7 Conclusion

Today when dealing with job crafting, we need to consider the different degrees of freedom, depending on which specific organization one works in, and which specific job one possesses. Workers today are far more responsible for their job satisfaction, motivation and future careers. If job crafting is to be a success, the individual possibility of exploiting benefits and available resources of the job is crucial. In addition, by focusing on job crafting one can influence the individual work engagement and organizational performance. One must also notice that job crafting also has a negative side. In some individuals, job crafting can have a negative effect on others when building relationships, achieving job resources, and determining the workload. Job crafting can have a negative effect on the organization, when the employee's job crafting goes against goals and values of the organization. In a coping perspective, increasing positive job crafting and decreasing all forms of negative crafting, become especially important. Based on this, leaders should have clear targets for job crafting. They should follow up on job crafting through appraisal interviews and follow-up calls.

Job crafting can be a useful tool in situations of change so that the workers sense that change is meaningful and successful (Saksvik & Thun, 2014). For example, with older employees, and with employees with special needs or chronic illnesses, job crafting can be especially relevant, as well as facilitating people who are planning to return from sick leave. Job crafting should not try to substitute for top-down initiatives or interventions, but it is an effective strategy for organizations and managements to take a closer look at the phenomenon, as this can promote motivation and job performance in a constantly changing work environment.

References

Airila, A., Hakanen, J. J., Schaufeli, W. B., Luukkonen, R., Punakallio, A., & Lusa, S. (2014). Are job and personal resources associated with work ability 10 years later? The mediating role of work engagement. *Work and Stress, 28*(1), 87–105. doi:10.1080/02678373.2013.872208.

Aust, B., Rugulies, R., Finken, A., & Jensen, C. (2010). When workplace interventions lead to negative effects: Learning from failures. *Scandinavian Journal of Public Health, 38*(3), 106–119. doi:10.1177/1403494809354362.

Bakker, A. B., & Demerouti, E. (2007). The job demands-resources model: State of the art. *Journal of Managerial Psychology, 22,* 309–328.

Bakker, A. B., & Demerouti, E. (2008). Towards a model of work engagement. *The Career Development International, 13*(3), 209–223.

Bakker, A. B., & Demerouti, E. (2014). Job demands-resources theory. In I P. Y. Chen, & C. L. Cooper (Eds.), *Work and wellbeing: wellbeing: A complete reference guide* (Vol. III; pp. 37–64). Chichester, UK: Wiley-Blackwell.

Bakker, A., Hakanen, J., Demerouti, E., & Xanthopoulou, D. (2007). Job resources boost work engagement, particularly when job demands are high. *Journal of Educational Psychology, 99* (2), 274–284.

Bakker, A. B., Tims, M., & Derks, D. (2012). Proactive personality and job performance: The role of job crafting and work engagement. *Human Relations, 65,* 1359–1378. doi:10.1177/0018726712453471.

Berg, J. M., Dutton, J. E., & Wrzesniewski, A. (2013). Job crafting and meaningful work. In: I B. J. Dik, Z. S. Byrne, & M. F. Steger (Eds.), *Purpose and meaning in the workplace* (pp. 81–104). Washington, DC: American Psychological Association.

Berg, J. M., Grant, A. M., & Johnson, V. (2010). When callings are calling: Crafting work and leisure in pursuit of unanswered occupational callings. *Organization Science, 21*(973), 994. doi:10.1287/orsc.1090.0497.

Christensen, M. (2008). *Positive factors at work—the first report of the Nordic project (TemaNord 2008:501).* Copenhagen: Nordic Council of Ministers.

Christensen, M., Aronsson, G., Clausen, T., Hakanen, J., & Vivoll Straume, L. (2012). *Building engagement and healthy organisations. A test of the Nordic questionnaire on Positive Organisational psychology (N-POP)* (p. 549). TemaNord.

Christian, M. S., Garza, A. S., & Slaughter, J. E. (2011). Work engagement: A quantitative review and test of its relations with task and contextual performance. *Personnel Psychology, 64,* 89–136. doi:10.1111/j.1744-6570.2010.01203.x.

Crawford, E. R., LePine, J. A., & Rich, B. L. (2010). Linking job demands and resources to employee engagement and burnout: A theoretical extension and meta-analytic test. *Journal of Applied Psychology, 95*(5), 834–848.

Demerouti, E. (2014). Design your own job through job crafting. *European Psychologist, 19*(4), 237–247. doi:10.1027/1016-9040/a000188.

Demerouti, E., Bakker, A. B., Nachreiner, F. & Schaufeli, W. B. (2001). The job demands-resources model of burnout. *Journal of Applied Psychology, 86,* 499–512. http://dx.doi.org/10.1037/0021-9010.86.3.499.

Ghitulescu, B. E. (2007). *Shaping tasks and relationships at work: Examining the antecedents and consequences of employee job crafting* (Doctoral dissertation). Retrieved from http://etd.library.pitt.edu/ETD/available/etd-12112006131255/unrestricted/ghitulescube_etd.pdf.

Hackman, J. R., & Oldham, G. R. (1980). *Work redesign.* Reading MA: Addison-Wesley.

Hakanen, J. J., & Schaufeli, W. B. (2012). Do burnout and work engagement predict depressive symptoms and life satisfaction? A three-wave seven-year prospective study. *Journal of Affective Disorders, 141,* 415–424. doi:10.1016/j.jad.2012.02.043.

Hakanen, J. J., Perhoniemi, R., & Toppinnen-Tanner, S. (2008). Positive gain spirals at work: From job resources to work engagement, personal initiative and work-unit innovativeness. *Journal of Vocational Behavior, 73,* 78–91. doi:10.1016/j.jvb.2008.01.003.

Hakanen, J. J., Bakker, A. B., & Demerouti, E. (2005). How dentists cope with their job demands and stay engaged: The moderating role of job resources. *European Journal of Oral Sciences, 113,* 479–487.

Halbesleben, R.B. (2010). A meta-analysis of work engagement: Relationships with burnout, demands, resources, and consequences. In I A. B. Bakker & M. P. Leiter (Eds.), *Work engagement: a handbook of essential theory and research.* New York: Psychology Press.

Halbesleben, J. R. B., & Wheeler, A. R. (2008). The relative roles of engagement and embeddedness in predicitng job performance and intention to leave. *Work and Stress, 22,* 242–256.

Herzberg, F. (1966). *Work and the nature of man.* Cleveland: World Publishing Company.

Hetland, H. (2008). Transformasjonsledelse: Inspirasjon til endring. [Transformational leadership: Inspiration for change]. *Tidsskrift for Norsk Psykologforening, 45*(3), 265–271.

Hobfoll, S. E. (2001). The influence of culture, community, and the nested-self in the stress process: Advancing conservation of resources theory. *Applied Psychology: An International Review, 50*(3), 337–421.

Karasek, R. (1979). Job demands, job decision latitude and mental strain: Implications for job redesign. *Administration Science Quarterly, 24*(2), 285–308.

Kira, M., Balkin, D. B., & San, E. (2012). Authentic work and organizational change: Longitudinal evidence from a merger. *Journal of Change Management, 12,* 31–51.

Kubota, K. A., Shimazu, A., Kawakami, N., Takahashi, M., Nakata, A., & Schaufeli, W. B. (2011). Association between workaholism and sleeping problems among hospital nurses'. *Industrial Health, 48,* 864–871.

Leana, C., Appelbaum, E., & Shevchuk, I. (2009). Work process and quality of care in early childhood education: The role of job crafting. *Academy of Management Journal, 52,* 1169–1192.

Locke, E. A., & Latham, G. P. (1990). *A theory of goal setting and task performance.* Englewood Cliffs: Prentice Hall.

Lyons, P. (2008). The crafting of jobs and individual differences. *Journal of Business and Psychology, 23,* 25–36. doi:10.1007/s10869-008-9080-2.

Mauno, S., Kinnunen, U., Mäkikangas, A., & Feldt, T. (2010). Job demands and resources as antecedents of work engagement: A qualitative review and directions for future research. In I S. Albrecht (Ed.), *Handbook of employee engagement* (pp. 111–129). Australia: Elgar publishing.

Petrou, P., Demerouti, E., Peeters, M., Schaufeli, W. B., & Hetland, J. (2012). Crafting a job on a daily basis: Contextual antecedents and the link to work engagement. *Journal of Organizational Behavior, 33,* 1120–1141.

Saksvik, P. Ø., & Thun, S. (2014). Ansvar for egen omstilling. [Responsibility for one's own adjustment process]. I P. Ø. Saksvik (Ed.). *Festskrift for Ivar A. Bjørgens 80-årsdag.* Trondheim: Psykologisk institutt, NTNU.

Salanova, M., Agut, S., & Peiró, J. M. (2005). Linking organizational resources and work engagement to employee performance and customer loyalty: The mediation of service climate. *Journal of Applied Psychology, 90*(6), 1217–1227.

Schaufeli, W. B., & Bakker, A. B. (2003). *The Utrecht Work Engagement Scale (UWES). Test manual.* Utrecht, The Netherlands: Department of Social & Organizational Psychology.

Schaufeli, W. B., & Bakker, A. B. (2004). Job demands, job resources, and their relationship with burnout and engagement: A multi-sample study. *Journal of Organizational Behavior, 25*(3), 293–315.

Schaufeli, W. B., & Bakker, A. B. (2010). The conceptualization and measurement of work engagement: A review. in I A.B. Bakker & M.P. Leiter (Eds.), *Work engagement: A handbook of essential theory and research.* New York: Psychology Press.

Schaufeli, W. B., Taris, T. W., & Van Rhenen, W. (2008). Workaholism, burnout, and work engagement: Three of a kind of three different kinds of employee well-being? *Applied Psychology: An International Review, 57,* 173–203.

Schaufeli, W. B., Bakker, A. B., & Van Rhenen, W. (2009). How changes in job demands and resources predict burnout, work engagement, and sickness absenteeism. *Journal of Organizational Behavior, 30,* 893–917.

Schaufeli, W. B., Salanova, M., Gonzalez-Roma, V., & Bakker, A. B. (2002). The measurement of burnout and engagement: A confirmatory factor analytic approach. *Journal of Happiness Studies, 3,* 71–92.

Siegrist, J. (1996). A diverse health effects of high-effort/low-reward conditions. *Journal of Occupational Health Psychology, 1*(1), 27–41.

Tims, M., & Bakker, A. B. (2010). Job crafting: Towards a new model of individual job redesign. *South African Journal of Industrial Psychology, 36,* 1–9.

Tims, M., Bakker, A. B., & Derks, D. (2012). Development and validation of the job crafting scale. *Journal of Vocational Behavior, 80,* 173–186. doi:10.1016/j.jvb.2011.05.009.

Vroom, V. H. (1964). *Work and motivation.* New York: Wiley.

Wrzesniewski, A., & Dutton, J. E. (2001). Crafting a job: Revisioning employees as active crafters of their work. *The Academy of Management Review, 26*(2), 179–201.

Wrzesniewski, A., LoBuglio, N., Dutton, J. E., & Berg, J. M. (2013). Job Crafting and Cultivating Positive Meaning and Identity in Work. In A.B. Bakker (Ed.), *Advances in positive organizational psychology* (pp. 281–302). Bingley: Emerald Group Publishing Limited.

Xanthopoulou, D., Bakker, A. B., Demerouti, E., & Schaufeli, W. B. (2009). Work engagement and financial returns: A diary study on the role of job and personal resources. *Journal of Occupational and Organizational Psychology, 82,* 183–200.

Chapter 8
Constructive Stress

Per Øystein Saksvik

There is a common view in the stress literature that there exists a curvilinear relationship between positive stress (eustress) and negative stress (distress). This model does not imply that eustress is at one end of the scale and distress at the other, but rather the model has a bell-shaped curve. At one end of the curve, less challenging tasks with low work demands cause distress. The top of the curve is where moderate work demands of moderately challenging tasks cause maximal eustress. At the other end of the curve, distress is characterized by tasks that are too challenging and work demands too high. This model, called the Yerkes–Dodson law, is one of the oldest in the history of psychology, dating back to 1908 (Fig. 8.1).

This view of stress as a phenomenon has lived on through Hans Selye, one of the big theorists on the subject. He began his work in the 1930s and did his most influential work in the 1950s. Selye coined the terms "eustress" and "distress" (Cooper & Payne, 1992). In recent research, the main focus has still been on how the two types of stress operate together and how they can merge to explain the total stress experienced by the individual. In a model introduced by Nelson and Simmons (2003), the experience of stress is tied to one and the same starting point (called a stressor), which can be performance demands in the workforce, which give some people negative experiences (distress) whilst others can grow through the challenges and experience satisfaction (eustress). Nelson and Simmons developed a complicated model to include both the positive and negative causes of stress. The model begins with a sum of stressors on the left-hand side, including everything from physical to psychological to relational aspects—for example, different role expectations or job promotions. The next box in the model is based on individual differences—for instance, optimism and resilience. This again can lead to the

P.Ø. Saksvik (✉)
Department of Psychology, Norwegian University of Science and Technology (NTNU), Trondheim, Norway
e-mail: per.saksvik@ntnu.no

M. Christensen et al. (eds.), *The Positive Side of Occupational Health Psychology*,
https://doi.org/10.1007/978-3-319-66781-2_8

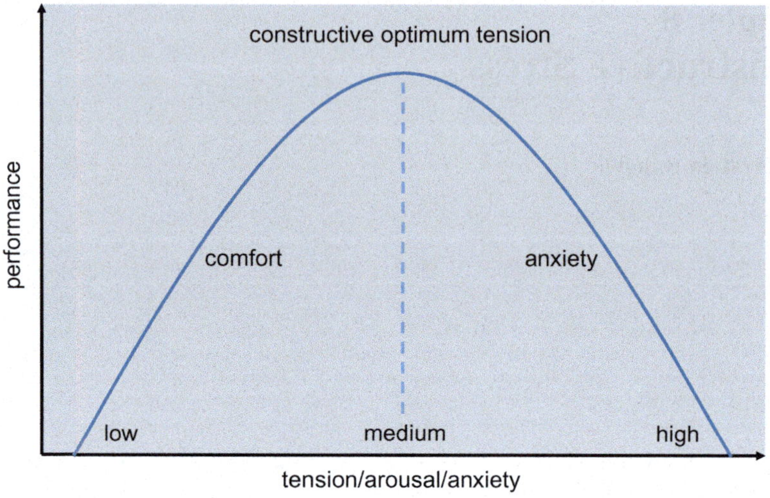

Fig. 8.1 The Yerkes–Dodson curve

experience of eustress or distress. There are two mechanisms that start the process of managing the stressors in this model, one called coping and the other savouring. They are applicable for managing positive and negative stress as well as themselves influencing the stressors—that is, to what degree the stressors are perceived as irksome.

The consequences of this experience will differ greatly depending on how the individual is able to cope with the stressors at hand. For some, the consequence will be to take sick leave, whilst others will heighten their work achievement. As stated earlier, this tends to happen because of different personalities. Research on normal workers from Saksvik, Saksvik, and Nordvik (2004) showed that those that scored high on neuroticism and openness experienced negative stress (distress) more easily than others, whilst those who scored high on extraversion to a larger extent experienced eustress.

A popular analogy of the relationship between eustress and distress is the bathtub model (Nelson & Cooper, 2005) (Fig. 8.2).

The two researchers claim that the bathtub can be used as an analogy for explaining the condition between distress and eustress. If you are taking a bath in a tub filled only with cold water, you will soon (or right away) experience discomfort. However, there are different regulating mechanisms in a bathtub, mechanisms like hot and cold water and the outflow. By regulating these three options, the researchers claim that it will give you a comfortable temperature in the bathtub. The analogy from Nelson and Cooper may not be exactly optimal. By pouring only hot water into the tub, you will soon find the same (even dangerous) discomfort. If the water is too cold or too hot, you may consider not getting in or pulling the plug and then refilling the bathtub with water that better fits your comfort. Additionally, there are those who enjoy ice bathing and find the cold water sensation refreshing. Personal differences

Fig. 8.2 The bathtub model. *Photo* Colourbox.com

are taken into consideration in this model, but there are those who think that this model does not give the whole picture of the relationship between eustress and distress. Maybe this indicates that there is no direct relationship?

The notion of there being parallel but different mechanisms that support distress and eustress is not a popular opinion in the literature, although it is a common assumption that good and bad can be perceived differently in people's lives. This way of thinking in terms of separation implies that negative and positive stresses live side by side but have different triggers, experiences and consequences, meaning that it is possible to live with both high negative stress and high positive stress in a competing battle. There is also the possibility that this competition is healthy, in that they stimulate one another, as seen when we compete in sports. It is normal to experience nervousness before a competition and it is known to make a positive contribution to the performance, even if it feels negative at first. As the race starts, the body releases energy that stimulates us to optimal performance and a feeling of happiness after passing the finish line. This feeling of happiness becomes stronger if the negative stress was great before and, to an extent, during the competition. In accordance with this thinking, we could not live without either the positive or the negative stress: it is the collaboration over time that yields the result. If we consider beyond the competition example, we can think of how this collaboration affects us: we experience a longer period of stable pressure and find ourselves on the limit of it, straining both body and mind, but the pressure is necessary to complete the task and to achieve the expected outcome. If we go back to the competition analogy, we see the amount of effort that is expended in a period where the athlete finds him/herself on the borderline between negative and positive effect and where the final result is dependent on this initial effort. This view can be understood as a balance between the forces, where the optimal result is when there is equality. In the literature, this balance perspective is mostly known through the popular JD-R model. On the one hand, the model includes the demands of work, including physical, psychological, social or organizational aspects of the job that require maintenance of physical and/or psychological effort and are in this way associated with physiological or psychological affordances (Bakker & Demerouti, 2007). On the other hand, the resources of the work represent the physical, psychological, social or organizational aspects that contribute to achieve the tasks: these contribute to a reduction in the demands of the job and the physiological and psychological demands associated with them, and/or stimulate personal growth, education and development (Bakker & Demerouti, 2007; see also Chaps. 3 and 4) (Fig. 8.3).

The benefit in respect of the competition (and not the balance) between these two forces is that the ideal is not stability but uneasiness. Instability allows progress to flourish, whilst stability gives a status quo and poorer prognoses (see Chap. 3).

Fig. 8.3 The balance perspective. *Photo* Colourbox.com

This is the core of chaos theory, which has been receiving acknowledgement in organization and leadership literature (Burns, 2004). The competition perspective has gathered inspiration from chaos theory and considers how an organization should grow and develop. The main idea of this theory is the complete opposite of the balance perspective. This view considers that stability leads to stagnation and, in the worst scenario, death (the organization experiences bankruptcy). Conversely, unbalance and disharmony lead to the embedded values being changed and drive the organization forward (Saksvik & Tvedt, 2009). Seen this way, maybe this view should rather be called "constructive conflict" (Tvedt, Saksvik, & Nytrø, 2009; see also Chap. 3).

The aim is to apply this theory both to individuals and to relationships between co-workers. Saksvik and Tvedt (2009) claim that the manager's wish is to ensure that the workers in his/her working group have access to the expected information concerning the organizational progress and future, as well as to ensure that the co-workers share values, all whilst inspiring them to challenge and change their values. The manager should support a culture of experimentation, be open and learn from mistakes or misunderstandings. This can afford his/her employees new ways of challenging the fundamental values that are embedded in the organization and develop insights into change that is unanticipated and, in some cases, irrational. One of the important reasons for organizations failing in achieving their goals is a lack of openness about their attitude towards change (Nytrø, Saksvik, Mikkelsen, Bohle, & Quinlan, 2000). In this way, it is a paradox that making mistakes is a requirement for the organization to start over.

Stimulating openness about mistakes is a term for learning from our mistakes and gets us thinking in new ways. An organization always moves towards stability according to its environment—that is to say, it must be capable of both single-loop and double-loop learning (Argyris & Schön, 1996). Organizations should not fear failure: such a fear hinders creativity and development.

This, though, is applicable to the individual as well as to the collective work force. We have previously discussed the old notion of the workers' collective from Lysgaard (1967) in a context of joint norms (Saksvik, Hammer, & Nytrø, 2013) (see Chap. 3). Collective control has to do with a common experience in the work environment. In agreement with Lysgaard, we emphasize the informal side of the social system regarding social norms in the workplace. This gives the idea that there is a common understanding amongst the workers and that they are well aware of this notion.

8.1 Collective Stress Coping

Lansisalmi, Peiro, and Kivimaki (2000) uncovered the collective strategies for coping with stress in a study of three dimensions of a multinational company. The strategies were uniform learned answers on tackling stressors; communal efforts to interpret a situation and joint effort towards tackling negative internal feelings. They concluded that the stress experience and the coping strategies have a collective quality. A culture develops that not only moderates the experience of stress, but also contains collective coping answers to stress factors that seem to have their origin in the organizational environment.

This is also true of collective management of negative stress. Clearly, there will be a collective dimension of managing positive stress as well, but this is probably not embedded in the concept of stress. In our article about collective control, we used the concept of norms. Norms can reflect both negative and positive values in an organization. The norms that are found in our scale of norms are, for example, "there is a mutual trust between the managers and the workers" and "we help each other with personal problems" (Hammer, Saksvik, Nytrø, Torvatn, & Bayazit, 2004). The norms can be understood from the social capital that subsists in an organization. Hasle and Møller (2007) demonstrated how a culture based on conflict and contradictory interests was of less service than a culture based on common faith and respect, which builds better social capital. Oksanen, Kouvonen, Kivimäki and Pentti (2008) pointed out that higher levels of social capital were associated with ensuring good health. In the same article from 2013, we made a division between relational self-identity, which is based on relationships between specific associates or managers, and collective self-identity, in which we define ourselves in relation to our group membership.

Apparently, it is important for us to act in accordance with a community. This gives rise to us managing the negative stress, but also to build social capital through identifying which positive values are important to us. In the same way that the difficult stress demands a lot of us and can be manageable through collective coping strategies, the positive stress will also be demanding. If we have clear norms on how to interact for the best of the community, this will make life easier. We need maps to show us where we should go.

8.2 Norms for Positive Interactions in the Workplace

Putnam defined social capital as the sum of norms, trust and network (Putnam, 1988), which is a comprehensive definition. Putnam's definition notes that the degree of trust in one's experience of justice in an organization is an expression of the company's collaboration. This is estimated by three relationships: bonding (within a unit or department), bridging (between units or departments) and linking (between the management and co-workers).

In one of our studies, we have found that increased self-knowledge before changes—i.e., being aware of your own norms and your own reaction patterns—is fundamental for succeeding in organizational change. In addition, it is noted that the presence of the leadership, the clarification of roles in terms of time and good management of conflict is important for adjustment (Saksvik et al., 2007).

We have developed our own scale of norms, consisting of shared understanding between co-workers in the workplace (Hammer et al., 2004). The notion of shared understanding means that a question should be asked as if there are more than just you that share this notion. This means that there is a common conception between several co-workers and all are familiar with this conception. A scale of norms naturally includes both positive and negative norms. When we focus in this chapter on positive stress, it is important to distinguish between the positive norms and the negative norms. We have yet to inquire into whether the positive and negative norm groups appear as independent factors in a factor analysis. Other separations may occur between norms that are more evident and groups into which they are categorized—for example, as norms for group cohesion or norms for productivity. It is interesting to see if there is a division between the positive and the negative norms: this can tell us more about what we started off with in this chapter—that is, whether the positive and the negative stress exist in parallel or if there is an interaction.

There is an implication that points towards the good stress being crucial on both an individual and a collective level. Stress is vital for an organization to work over time, but we need norms on how to interact to make use of the stress. It is probably a subtle balance between when the stress is constructive and when it is destructive. This does not mean that the stress either uplifts or destroys our coexistence: it means that we have to live with both of them, but get better at shifting the perspective from the negative to the positive. Constructive stress gives rise to growth and development through us working better together in an organization. Still, counterforces and challenges are something we need in order to develop our understanding. When we master hardship, we can progress.

References

Argyris, C., & Schön, D. A. (1996). *Organizational learning II.* USA: Addison-Wesley Publishing Company.

Bakker, A. B., & Demerouti, E. (2007). The job demands-resources model: State of the art. *Journal of Managerial Psychology, 22*(3), 309–328. doi:10.1108/02683940710733115.

Burns, J. S. (2004). Chaos theory and leadership studies: Exploring uncharted seas. *Journal of Leadership & Organizational Studies, 9,* 42–56.

Cooper, C. L., & Payne, R. (1992). International perspectives on research into work, well-being, and stress management. In J. C. Quick, L. R. Murphy, & J. J. Hurrell (Eds.), *Stress and well-being at work (ss* (pp. 348–368). Washington, DC: American Psychological Association.

Hammer, T. H., Saksvik, P. Ø., Nytro, K., Torvatn, H., & Bayazit, M. (2004). Expanding the psychosocial work environment: Workplace norms and work-family conflict as correlates of stress and health. *Journal of Occupational Health Psychology, 9,* 83–97.

Hasle, P., & Møller, N. (2007). From conflict to shared development: Social capital in a Tayloristic environment. *Economic and Industrial Democracy, 3,* 401–429.

Lansisalmi, H., Peiro, J. M., & Kivimaki, M., IV. (2000). Collective stress and coping in the context of organizational culture. *European Journal of Work and Organizational Psychology, 9*(4), 527–559. doi:10.1080/13594320050203120.

Nelson, D., & Cooper, C. (2005). Stress and health: A positive direction. *Stress and Health, 21,* 73–75. doi:10.1002/smi.1053

Nelson, D. L., & Simmons, B. L. (2003). Health psychology and work stress: A more positive approach. In J. C. Quick & L. E. Tetrick (Eds.), *Handbook of occupational health psychology* (ss. 97–118). Washington, DC: American Psychological Association.

Nytrø, K., Saksvik, P. Ø., Mikkelsen, A., Bohle, P., & Quinlan, M. (2000). An appraisal of key factors in the implementation of occupational stress interventions. *Work & Stress, 13,* 213–225.

Oksanen, T., Kouvonen, A., Kivimäki, M., & Pentti, J. (2008). Social capital at work as a predictor of employee health: multilevel evidence from work units in Finland. *Social Science and Medicine, 3,* 637–649.

Putnam, R. D. (1988). Diplomacy and domestic politics: The logic of two-level games. *International Organization, 42*(3), 427–460.

Saksvik, I. B., Saksvik, P. Ø., & Nordvik, H. (2004). Personlighetens rolle for hvordan stress oppleves i arbeidslivet. *Tidsskrift for Norsk Psykologforening, 41,* 612–619.

Saksvik, P. Ø., & Tvedt, S. D. (2009). Leading change in a healthy way. *Scandinavian journal of Organizational Psychology, 1,* 20–29.

Saksvik, P. Ø., Tvedt, S. D., Nytrø, K., Buvik, M. P., Andersen, G. R., Andersen, T. K., et al. (2007). Developing criteria for healthy organizational change. *Work & Stress., 21,* 243–263.

Tvedt, S. D., Saksvik, P. Ø., & Nytrø, K. (2009). Healthy organizational change processes as a prerequisite for a good psychological work environment. *Work & Stress, 23,* 80–98.

Chapter 9
Coworkership and Prolific Behaviors in Modern Work Life

Martin Schrøder, Marit Christensen, Siw Tone Innstrand and Anette Fjeld

There is a consensus in most organizations that in order to develop a healthy and productive work environment, you have to arrange for a good job design that promotes engagement and positive health and have a competent, good leader. This creates a good basis for a sound and sustainable environment. But a good working environment is not something you get, but something you actively take part in creating through your own efforts and in relation to your managers and colleagues. The collective effort from both the organization and the coworkers is essential for a healthy and productive organization. The coworkers' individual role in organizational development is often somewhat overlooked; however, in this chapter, we argue that coworkers individually play an active and important part in creating a sustainable work place.

Organizational psychologists have for a long time argued about what constitutes a good employee. Consequently, researchers have presented multiple constructs to reflect employee behavior that goes beyond any requirement or expectation of supervisors and the organizational conditions that promote such behavior. The concept of coworkership is a rather new construct that pertains to such work behavior and has received growing recognition in Scandinavian labor studies during this past decade (Kilhammar & Ellström, 2015). Research on coworkership has so far been limited and there is no clear consensus on the concept's meaning and content. The term has thus been defined in several different ways. With its novelty and focus on employees' relationships in the workplace (Tengblad, 2010), it is essentially a contemporary human resource development strategy that attempts to

M. Schrøder (✉) · M. Christensen · A. Fjeld
Department of Psychology, Norwegian University of Science and Technology (NTNU), Trondheim, Norway
e-mail: martin.schroder89@gmail.com

S.T. Innstrand
Department of Public Health and Nursing, Faculty of Medicineand Health Sciences, Norwegian University of Science and Technology, Trondheim, Norway

M. Christensen et al. (eds.), *The Positive Side of Occupational Health Psychology*,
https://doi.org/10.1007/978-3-319-66781-2_9

improve employees' responsibility and performance, while creating job satisfaction and participation (Kilhammar & Ellström, 2015). The types of employee behaviors that are posited to follow this strategy, however, seem to overlap with several other concepts (Kilhammar & Ellström, 2015), yet it is suggested that they differ in several aspects (Andersson & Tengblad, 2007).

The aim of this chapter is therefore to explore the novel concept of coworkership and how it relates to or differs from similar constructs, and finally to discuss its theoretical and practical implications.

9.1 Coworkership

The concept of coworkership was developed in a Scandinavian context, or more precisely, in the context of the Swedish labor market (Kilhammar & Ellström, 2015). Tengblad (2010) argued that coworkership summarized the Scandinavian organizational philosophy which relies on the relationship between workers and their work, managers, and colleagues alike. This philosophy states that employees should take on responsibilities, be active, and use the possibilities within the organization to learn and develop through cooperation with others, and take on challenging tasks. This line of thought, then, is the result of the last 40 years of labor market evolvement and is a representation of a working life that unites effectiveness, performance, work engagement, and well-being (Tengblad, 2010). In their pioneering work, Andersson and Tengblad (2007) argued that coworkership is a descriptive, as well as a normative concept of employee behavior.

9.2 Coworkership as a Descriptive Concept

As a descriptive concept, coworkership is defined as "how employed personnel handle the relation to their employer, their colleagues, and their own work" (Andersson & Tengblad, 2007). The descriptive concept coworkership, then, is one concerning people's relationships in organizations and the practices within these relationships. The behaviors that are associated with the definition, according to Andersson and Tengblad (2007), are those that represent loyalty, responsibility, initiative, and cooperation. Tengblad (2010) presented two points of view on coworkership (see Table 9.1) to better understand the concept. The first view was that coworkership could take on different roles in an organization. Consequently, Tengblad (2010) posited five different coworkership roles associated with work in organizations. The first role, traditional, refers to coworkership in organizations that is *not* characterized by active employees. Such a role is traditional in the sense that managers assign tasks and directions to the workers conducting the task. This type of role is more likely to be found in retail and storage occupations than other lines of work. The second role, organization-oriented, can be found in businesses where active and responsible work roles have been implemented for employees. There is, however, an expectation of how each work role will be carried out, and because of this, the roles are not completely autonomous. Rather, responsibilities are delegated to employees who are expected to produce certain results. The third role, group-oriented, entails an organization in which work groups have taken all, or a big portion, of the manager's function. In these cases, the supervisor is included in the work group but is not a manager per se. Rather, the work group has a great influence on and responsibility for how the work is done and by whom the tasks will be completed. Fourth, coworkership can be recognized as individual oriented when employees, as individuals, act independently. This is done when employees take the initiative and make decisions without being directed by supervisors. In the individual-oriented coworkership role, employees are often engaged in their work and directed in their personal development during the completion of tasks. However, personal development is often in the context of their own work and their sense of self, rather than the development of organizational conditions. This also presents the possibility of an emerging competitive climate, in which coworkers view each other as competitors rather than colleagues (Tengblad, 2010); this can be detrimental to team efforts. The last coworkership role is called leaderless and is a role with an undeveloped leader–member relationship. This means that the leader has a peripheral role in the employees' work, and that leaders and employees do not work directly with one another. This is more typical in more autonomous lines of work, like dental care, psychiatric work, and judicial work. Leaderless coworkership can be good when employees take a larger share of responsibility for their own work and learn to work independently. However, it can also create a larger gap between leaders and employees in cases where cooperation and mutual obligations are needed, or asked for.

The second way to describe coworkership is by the employees' skills and attitudes toward their work. Tengblad (2010) suggested four categories that described these factors as forms of coworkership: compliant, specialized, proactive, and boundaryless (see Table 9.1). The first category, compliant coworkership, is more usual when employees are new to the organization, or have just started a new job. As expectations of newcomers are to follow the routines and instructions of the organization, he/she tries to adapt to the organizational culture and systems, at the same time that tasks are learned and relationships are formed. This form of coworkership, however, is passive as it is about following orders and imitating the behaviors of others. The employee, then, must engage in other types of coworkership to experience work development. The second category, specialized

Table 9.1 Description of types of coworkership

Coworkership	
Organizational coworkership roles	Individual development level
Traditional	Compliant
Organization-oriented	Specialized
Group-oriented	Proactive
Individual-oriented	Boundaryless
Leaderless	

coworkership, refers to employees who have a particular responsibility for completing tasks, or a set of skills needed to complete tasks, whether specified by the management, or unexpressed by any form of guideline. In this type of coworkership, employees are trained to take responsibility for the defined tasks and are given the opportunity to later become experts. There are, however, potential pitfalls to having a lot of experts in an organization. When employees master their own area of expertise, communication problems can arise. Also, if confronted with change, employees can experience conservatism as altering their preferred way of working can result in (temporary) inefficiency. The third category, proactive coworkership, refers to employees who have developed an active approach to their own and others' work tasks, and to the betterment of the organization as a whole. Proactive coworkership can thus be a source of routine improvements, cooperation, and exchange of information. The employees are responsible for their own work, and furthermore, are proactive in taking responsibility. The last category is called boundaryless coworkership. During boundaryless coworkership, employees are so good at creating an efficient work place that they become an integral part of the organizational collective structure (Tengblad, 2010). They partake in decision-making processes and help each other when problems arise. Creating this category of coworkership in organizations is demanding, but the rewards are hypothesized to be a high-yielding business with employees experiencing high levels of job satisfaction.

9.3 Coworkership as a Normative Concept

Coworkership as a normative concept reflect an organizational ideal in which employees are respected, motivated, collaborative and active (Andersson & Tengblad, 2007, p. 3). For this organizational ideal to exist, four antecedents have been proposed based on previous research (Andersson & Tengblad, 2007). The antecedents are related to managers and employees alike and reflect their relationship in which they take responsibility for their actions, attitudes, and the relationship as a whole. The four antecedents are (1) trust and openness, (2) fellowship and cooperation, (3) commitment and meaningfulness, and (4) accountability and agency (Andersson & Tengblad, 2007). Trust and openness refers to the extent to which coworkers have open and honest communication. They trust each other, respect each other's opinions, and solve conflicts as they arise. Fellowship and cooperation reflect employees' sense of belonging in the work place, and their ability to cooperate across organizational structures, functions and professions, and the strengthening of relationships through such activities. Commitment and meaningfulness represents the degree to which employees find meaning in their tasks and work as a whole, and consequently strive to perform at work. The last antecedent, accountability and agency, refers to the employees' ability to take responsibility for their own work, the results of their work, and their willingness to develop their work capacities. Should these antecedents be present in an organization, it is posited that coworkership can evolve (Andersson & Tengblad, 2007).

In sum, coworkership is regarded as how employed personnel handle the relation to their employer, their colleagues, and their own work (in accordance with Andersson & Tengblad, 2007). Further, we will explore related concepts like the psychological contract, personal initiative, proactive behavior, organizational citizenship behavior, followership, and extra-role behavior to see how they are associated with coworkership.

9.4 Coworkership and Its Associates/Related Concepts

9.4.1 The Psychological Contract of Employees

The concept of psychological contracts was first introduced and given a place in organizational cultural studies by Argyris (1960), before it caught wind with the pioneering work of Rousseau (Rousseau, 1989; Skogstad, 2005). Rousseau (1989) defined the psychological contract as "an individual's beliefs regarding the terms and conditions of a reciprocal exchange agreement between that focal person and another party" (p. 123). In accordance with that notion, a psychological contract exists as an individual's perception of a reciprocal relationship of exchange between him/herself and the organization. Psychological contract theory, then, relies heavily on the notions of exchange and fairness that are posited by equity theory. In the context of psychological contracts, equity theory states that inequity/injustice exists for an employee when he/she perceives his/her input in the relationship with the organization to be greater than the input of the organization (Adams, 1965). For the continued existence of a psychological contract, it is therefore important that the relationship between the employee and organization is experienced as balanced.

An employee's experience of a psychological contract occurs when the employee perceives implicit or explicit promises of commitment from the organization (Robinson, 1996). However, these promises can also be the result of the individual's attributions of the underlying causes of their own behavior (Rousseau, 1989). For example, an employee might be under the impression that by exceeding the expectations of job performance, this act will lead to a pay raise, even though that may never have been stated by the organization. Also, Rousseau (1989) pointed out that organizations are only the context providers for the employees' creation of a psychological contract, and that they cannot have a psychological contract of their own. Thus, psychological contracts in the workplace can be regarded as different from the social norms of reciprocity, whereby people help those who have been helpful to them (Gouldner, 1960). If the organization fail to carry out what is perceived to be its obligation to the employee, the employee might experience a breach of the psychological contract. Since an employee believes that contributions to the organization will be reciprocated, trust is a core factor in the psychological contract (Rousseau, 1989). By not adhering to contract terms, a breach of the contract can consequently be regarded as a breach of the employee's trust in the organization, thus damaging the relationship between the parties. This perception, though, is a subjective experience and can occur even in the absence of an actual or deliberate contract breach (Robinson, 1996).

According to these notions, the psychological contract is not an employee behavior per se. Rather, it can be regarded as an antecedent or a precursor to behavior employees believe is expected by the organization. This is in line with coworkership as a normative concept. Since an employee with a psychological contract expects a reciprocal exchange relationship with the organization, he/she adapts his/her behavior to fulfill what is believed to be the obligation to the organization. Hence, as long as the employee perceives his/her input to be reciprocated, behavior that is believed to benefit the organization will continue. As the psychological contract is a matter of subjectivity (Robinson, 1996), what constitutes the desired behavior is mediated by the employees' perception of the contract conditions. Since the psychological contract is concerned with the subjective relationships between coworker and leader, it can therefore be considered an antecedent of coworkership.

9.4.2 Organizational Citizenship Behavior

Since the construct of organizational citizenship behavior (OCB) was first coined in the early 1980s, it has received growing attention in the fields of industrial and organizational psychology (LePine, Erez, & Johnson, 2002; Podsakoff, Whiting, Podsakoff, & Blume, 2009). Originally, OCB was defined as "individual behavior that is discretionary, not directly or explicitly recognized by the formal reward system, and that in the aggregate promotes the effective functioning of the organization" (Organ, 1988, p. 4). This definition, however, seems too semantically ambiguous when compared to other behavioral constructs, such as contextual performance (LePine et al., 2002). Goodman and Svyantek (1999) considered the contextual performance construct to include "(…)such activities as volunteering to carry out actions that are not formally a part of the job; helping others; following organizational rules procedures when personally inconvenient; endorsing and supporting organizational objectives; and persisting with extra effort to successfully complete one's task activities" (p. 255). Organ (1988), however, recognized the ambiguity of OCB and asserted that it was rather a discretionary and unrewarded behavior, thereby transcending contextual performance. Despite this, he later changed the definition of OCB so that the construct reflected "performance that supports the social and psychological environment in which task performance takes place" (Organ 1997, p. 95). Regardless of the continued similarities between the constructs, Organ's (1997) definition of OCB stands today as the one of the construct's most used definitions.

Regardless of the overlap with other construct definitions, researchers seem to agree that OCB is multidimensional. Several of these researchers have developed taxonomies for OCB, and though they share many similarities, Organ's (1988, 1990) taxonomy has been the most subject to rigorous research (LePine et al., 2002). Organ proposed a model with five dimensions: altruism, conscientiousness, sportsmanship, courtesy, and civic virtue. The dimensions were later conceptualized by Podsakoff, MacKenzie, Moorman, and Fetter (1990), where altruism was viewed as "behaviors that have the effect of helping a specific other person with an organizationally relevant task or problem" (p. 115). Conscientiousness was regarded as "behaviors on the part of the employee that go well beyond the minimum role requirements of the organization, in the areas of attendance, obeying rules and regulations, taking breaks, and so forth" (p. 115). Sportsmanship, on the other hand, was the "willingness of the employee to tolerate less than ideal circumstances without complaining" (p. 115). And as courtesy was regarded as "behavior on the part of an individual aimed at preventing work-related problems" (p. 115), civic virtue was "behavior on the part of an individual that indicates that he/she responsibly participates in, is involved in, or is concerned about the life of the company" (p. 115).

The concept of OCB is today regarded as discretionary behaviors that falls outside the conventional reward systems. OCB is posited to be positive for the social and psychological environment at work, and for the ease of research, conceptualized as a taxonomy. OCB is best compared to coworkership when described as a trait. While OCB is something that is above the norm, normal behavior is characterized by transactional relationships. It is therefore likely that the dimensions of altruism, courtesy, and the civic virtue of OCB could be part of the coworkership construct. However, because it is not stated that coworkership behaviors are recognized by any reward system or constrained by role requirements, one can also argue it is a form of OCB.

9.4.3 Personal Initiative

As working life increasingly changed, the concept of personal initiative (PI) was considered to increase organizational effectiveness and was introduced for inclusion in assessment centers (Frese, Fay, Hilburger, & Tag, 1997). Though several researchers had compared the dimensionality of individual performance with other constructs, Frese, Kring, Soose, and Zempel (1996) stated that previous constructs implied a lack of planning in the behavior. PI, on the other hand, was a self-starting, active approach to overcome difficulties in pursuing a goal. Thus, PI behavior could be the anticipation of and preparation for potential demands and problems that could arise in completing tasks. Thus, PI was defined as "a behavior syndrome resulting in an individual taking an active and self-starting approach to work and going beyond what is formally required in a given job" (Frese et al., 1996, p. 38). In the context of PI, a syndrome is when co-occurring behaviors put together signify initiative. Consequently, employees that go beyond formal role requirements are displaying PI behavior. There are a few characteristics of these behaviors: (1) the behavior is consistent with the organization's mission, (2) the behavior has a long-term focus, (3) the behavior is goal directed and action oriented, (4) the behavior persists in spite of barriers and setbacks, and (5) the behavior is self-starting and proactive (Frese et al., 1996).

Previous scholars have pointed out three basic tenets that stem from the definition of PI. The first tenet is that PI is self-starting. The use of the term self-starting in the definition implies that employees initiate behaviors without being told or being given explicit instructions, or that the action is due to role requirements (Frese & Fay, 2001). To achieve this requires goal setting, and it involves either a personal developed idea or from putting existing ideas or projects into action. The second tenet is that PI is proactive. Proactivity refers to having a long-term focus that enables employees to consider future demands and opportunities, and to actively do something about them. The future orientation helps employees to anticipate and handle their problems or to take advantage of contextual opportunities. Third, persistence is a necessity to reach goals. Because PI implies that change has occurred, this change is also in need of personal adaption on the part of the employee. Given that people usually do not embrace change, the person that takes the initiative needs persistence in order to break past these barriers. The tenets of PI are thought to reinforce each other, as a proactive future orientation makes it more likely that employees will develop goals that go beyond what is expected (Frese & Fay, 2001).

On the basis of the three basic tenets and from an action theory perspective, PI is thought to have three facets (Frese & Fay, 2001). First though, action theory states that people are active by nature, and that actions have a set of sequences—an action sequence (Frese & Zapf, 1994). An action sequence has four stages: (1) developing a goal that one wants to achieve, (2) collecting information and making a prognosis of future states, (3) information from earlier stages is used to create a plan that is executed later, and (4) during the execution, actions are monitored and feedback is gathered to adjust the actions (Frese & Fay, 2001). With the tenets, these four stages constitute the facets of PI, which are thereby a reflection of the tenets' manifestation at each stage of an action sequence.

In sum, PI is an employees' self-start of co-occurring behaviors that stretches past the formalities of a job description, in pursuit of a long-term goal that will benefit the organization. Accordingly, PI theory states that the behavior is self-starting, proactive, and persistent. With this in mind, researchers use action theory to describe how PI is manifested at different stages of behavior. Coworkership differs from personal initiative in the way it focuses on relationships; coworkership entails an element of proactivity because of the intention (future orientation) to handle relationships.

9.4.4 Proactive Work Behavior

Proactive work behavior (PWB) can be described as an umbrella construct that refers to specific employee behaviors that are self-starting and done on employees' own initiative (Parker, William, & Turner, 2006). Although proactive behaviors have been conceptualized at different levels (e.g., Ashford & Tsui, 1991; Kickul & Gundry, 2002; Simard & Marchand, 1995), here we are interested in the individual level. Of other employee behaviors presented in this chapter, personal initiative is probably the most closely related construct to PWB. But whereas personal initiative is directed toward future problems or opportunities, PWB focuses less on the future, and more on what can be improved in the present situation (Crant, 2000). Thus, Crant (2000) defined PWB as "taking initiative in improving current circumstances or creating new ones; it involves challenging the status quo rather than passively adapting to present conditions" (p. 436).

Due to the overlap with other constructs, PWB has also been described as a type of contextual performance (Morrison & Phelps, 1999). Researchers, however, argue that employees are able to be proactive and participative in both tasks and contextual elements, thereby not restricting proactive behavior to any specific context. Because of this, some researchers view PWB as a two-dimensional construct (Parker et al., 2006). The first dimension is called proactive idea implementation. This dimension refers to an employee taking the lead in executing an idea to improve the work place. The employee can do this by either implementing the idea themselves, or by voicing ideas to coworkers to start the implementation process. Frese, Teng, and Wijnen (1999) noted that the key to this dimension is that employees have sufficient autonomy to be able to implement ideas, and therefore, to be able to go straight to the implementation process rather than voicing the idea beforehand. The second dimension is proactive problem solving. This dimension involve self-starting responses to a problem. However, what constitutes a problem in one environment may be a routine in another. Thus, behaviors that are proactive in one organization may not be in others (Parker et al., 2006).

In short, PWB is behavior directed at creating new or improved circumstances in the organization, and consequently to drive the organization forward. Proactive behavior is recognized by being set forth by an idea of an employee, which is unusual or nonstandard for that particular work context. Note that proactive behavior is not necessarily an innovative process, as these constructs are more stringently defined in terms of novelty and the utility of the behavioral outcome (Parker et al., 2006).

Just like personal initiative, proactive work behavior is also seen as a self-starting initiative directed to drive the organization forward (Parker et al., 2006). However, compared to personal initiative which is directed toward the future, proactive work behavior is more concerned with the present situation (Crant, 2000).

9.4.5 Followership

When the academic focus on leadership held the prominent place in literature, the topic of followership remained understudied (Bjugstad, Thach, Thompson, & Morris, 2006). While leader and leadership studies were glorified, it excluded the larger group of workers: the employees. The article of Kelley (1988) therefore set the standard for thinking that success was not solely dependent on dynamic leaders leading passive subordinates, but that active followers also played a crucial role in organizational successes.

Though research on followership has picked up in recent years, it seems that no commonly accepted definition has emerged (Crossman & Crossman, 2011). In defining followership, Hollander and Webb (1955) pointed out that definitions could vary, depending on whether followership was approached from the perspective of a leader or a follower. Consequently, the definitions of followership are often related to the hierarchical differences between leaders and followers. The notion of hierarchical differences, however, is challenged by the more relational process between leaders and followers (Crossman & Crossman, 2011). As noted by Carsten, Uhl-Bien, West, Patera, and McGregor (2010).

> "Followership behaviors differ (from leadership) in that they do not address independent activities of occupying "subordinate" positions but behaviors of individuals acting in relation to a leader(s). In other words, followership behaviors are not about how individuals interact relative to their work or other coworkers but relative to those with higher status—with respect to leaders" (p. 545).

Consequently, followership is viewed as the complementing behavior of the employee to the current leadership. The construct of followership is therefore defined as "a relational role in which followers have the ability to influence leaders and contribute to the improvement and attainment of group and organizational objectives" (Crossman & Crossman, 2011, p. 484).

In a review of followership, Baker (2007) pointed out that the published literature on the subject had several different themes. Among the themes was that of a difference in leader and follower roles, as roles did not necessarily depict employee characteristics of individuals. Thus, employee behavior may be what is believed to be appropriate to the role of an employee and vice versa. Another theme that emerged from the literature was that leaders and followers shared a common purpose. For the relationship between leaders and followers to be fruitful, both groups need a common purpose on which to focus their work. By being interdependent, the leader needs to engage followers in mutually satisfying endeavors so interconnectedness and responsibility for meeting organizational goals are shared. These themes highlight the previous notion of a difference between leader and followers despite the possibility of similar personality characteristics and the interdependent relationship between follower and leader that is necessary for followership behaviors.

In categorizing followers according to dimensions of thinking and acting, Kelley (1992) developed a descriptive typology of followers. The thinking dimension consisted of critical, independent thinking at one end of the spectrum, and dependent, uncritical thinking at the other end. The acting dimension pertained to the type of engagement followers displayed in the organization as active or passive, that is, whether it was an active engagement creating positive energy, or a passive involvement creating negative energy. With these dimensions, Kelley (1992) asserted that there were five basic styles of followership.

To sum up, followership is a relational role where employees are seen to be able to influence leaders to reach group/organizational objectives. Often posited as a difference in the roles of leaders and followers, followership concerns the relationship between leaders and employees, and how the latter interacts with the former. Based on this, Kelley (1992) divided followership behavior into five different styles of followership based on the followers' display of independent thinking and type of engagement at work. Compared to coworkership, followership is a counterpart to leadership. The concept is oriented around leadership support and the making of effective leaders where the employees have a more passive role.

9.4.6 Extra-Role Behavior

As this chapter is finding, there are several constructs pertaining to which employee behaviors are seen as "good" or "desired." The case of extra-role behavior underscores this findings. An important point to make in this context is that there must be a difference between what is the expected behavior from an employee, and what behavior goes beyond this expectation. The behavior that is displayed in line with what is required and expected of a regular and ongoing job performance is referred to as in-role behavior (Katz, 1964). In cases where employees fail to deliver the required behaviors, they may also fail to receive organizational rewards, such as bonuses and pay increases, and consequently become subject to the disdain of supervisors. In contrast, supervisors value extra-role behaviors, as they are positive. Extra-role behaviors, then, are discretionary, unspecified by role descriptions, unrecognized by formal reward systems, and are not a source of punishment when they are not performed (Van Dyne & LePine, 1998). With these notions in mind, extra-role behavior is here defined as "those behaviors that go beyond specified role requirements, and are directed toward the individual, the group, or the organization as a unit, in order to promote organizational goals" (Somech & Drach-Zahavy, 2000, p. 650).

Somech and Drach-Zahavy (2000) noted that the definition of extra-role behavior stresses three main features of the construct. First, the behavior is voluntary; it is not part of the individual's job description, orders, or prescriptions from superiors, nor is it part of the job's role requirements. Second, behavior that is described as extra-role, is beneficial from an organizational perspective. Third, the definition highlights the multidimensionality of the construct. Because of inconsistent findings on extra-role behavior (George, 1996), there is a necessity for research to focus on all three levels (individual, team, and organization).

Based on the works of Katz (1964) and Organ (1988), Van Dyne, Graham, and Dienesch (1994) extended and categorized different organizational behaviors to a typology of extra-role behavior. The typology differentiated promotive behaviors from prohibitive behaviors, affiliative behaviors, and challenging behaviors. Promotive behaviors are proactive and cause things to happen, whereas prohibitive behaviors are more protective of those with less power and preventive by speaking out to stop destructive or inappropriate behavior. Affiliative behaviors, on the other hand, are behaviors that strengthen relationships and cooperation. Lastly, challenging behaviors emphasize ideas and current issues and can be disruptive to the status quo. Although this type of behavior can help the organization move forward, it can also damage relationships as change is not always wanted. This differentiation of behaviors can be used in numerous ways in research, i.e., to measure the more general type of behaviors that occur in a work place. Extra-role behavior, then, is work-related behavior that surpasses the expectations of a person's role requirements and supervisors, and that is not recognized by any reward systems. Because the focus of coworkership is on the handling of relationships, these behaviors could be described as a type of affiliative behavior of extra-role behavior. Therefore, coworkership has both similarities and differences to previous employee constructs.

9.5 Coworkership—Old Wine in New Bottles?

As noted earlier, coworkership as a descriptive concept is "how employed personnel handle the relation to their employer, their colleagues, and their own work (Andersson & Tengblad, 2007). Also, it has been posited that the factors of trust and openness, fellowship and cooperation, commitment and meaningfulness, and accountability and agency are necessary for coworkership (Andersson & Tengblad, 2007). These qualities of coworkership present some similarities to the constructs presented earlier in this chapter. First, there is an implicit notion of an existing relationship between the employee and the other party when coworkership behaviors are engaged in. This implies that there is a psychological contract mediating behaviors directed at strengthening the relationship with the employer. Second, it is not unlikely that the dimensions of altruism, courtesy, and the civic virtue of OCB could be a part of the coworkership construct. Though it is not stated that coworkership behaviors are explicitly recognized by any reward system, the anchors of coworkership could go beyond role requirements. Third, though different from PI in the way it focuses on relationships, coworkership entails an element of proactivity because of the intention (future orientation) to handle relationships. Fourth, because the focus of coworkership is the handling of relationships, these behaviors could be described as a type of affiliative behavior of extra-role behavior. Coworkership, then, has both similarities and differences to previous employee constructs. Even though these concepts touch upon some essential aspects of coworkership (i.e., responsibility, leader–employee relationships, reciprocal relations, and norm-based organizational behavior), coworkership highlights the worker's role in the organization in general and is seen as a more normal behavior related to a broader set of relationships (Kilhammar & Ellström, 2015).

Even though coworkership has been viewed as a Nordic term, it seems to exist in all workplaces regardless of culture. In a qualitative thesis study, similar coworkership values were identified among knowledge workers both in Norway and Hawaii (Fjeld, 2015). We therefore argue that the relevance and applicability of coworkership are broader. Although there is a growing interest in the concept of coworkership, and it is frequently used by HR practitioners, there is still a lack of measurement; a sound measurement is highly significant in order to study the antecedents and consequences.

For example, Andersson and Tengblad (2007) have posited that increased employability and career prospects, greater influence on one's own work (autonomy) and workplace decisions, and higher self-confidence and life satisfaction, were some of the possible consequences for individuals who developed good coworkership (p. 13). There is also the argument that individual factors, such as job satisfaction and perceived peer support, could be consequences at the individual level. When it comes to antecedents, it is likely that individuals who display high degrees of coworkership experience autonomy. Andersson and Tengblad (2007) argued that this notion of autonomous work was central to the emergence of coworkership in any workplace. Considering that individuals may not engage in behaviors that would benefit work relationships, organizational commitment could be an antecedent as well. More research on the antecedents and outcomes of coworkership is important to target interventions related to promoting coworkership. When it comes to a theoretical framework, coworkership could be founded on numerous theories of social exchanges, needs, motivation, psychological contracts, and so on.

We suggest that coworkership is an important concept to include in employee surveys as it offers a missing aspect of what constitutes a sustainable working environment. In conclusion, in organizational development an effort is needed from both the employee and the employer in order to create a healthy and productive psychosocial work environment.

References

Adams, J. S. (1965). Inequity in social exchange. *Advances in Experimental Social Psychology, 2*, 267–299.

Andersson, T., & Tengblad, S. (2007). *Analyzing concepts for work-life transformation—Making the case for coworkership*. Paper presented at the NFF-conference, August, 9th–11th, 2007, Bergen.

Argyris, C. (1960). *Understanding organizational behavior*. Oxford: Dorsey.

Ashford, S. J., & Tsui, A. S. (1991). Self-regulation for managerial effectiveness: The role of active feedback seeking. *Academy of Management Journal, 34*(2), 251–280.

Baker, S. D. (2007). Followership: The theoretical foundation of a contemporary construct. *Journal of Leadership and Organizational Studies, 14*(1), 50–60.

Bjugstad, K., Thach, E. C., Thompson, K. J., & Morris, A. (2006). A fresh look at followership: A model for matching followership and leadership styles. *Journal of Behavioral and Applied Management, 7*(3), 304–319.

Carsten, M. K., Uhl-Bien, M., West, B. J., Patera, J. L., & McGregor, R. (2010). Exploring social constructions of followership: A qualitative study. *The Leadership Quarterly, 21,* 543–562.

Crant, J. M. (2000). Proactive behavior in organizations. *Journal of Management, 26*(3), 435–462.

Crossman, B., & Crossman, J. (2011). Conceptualising followership—A review of the literature. *Leadership, 7*(4), 481–497.

Fjeld, A. (2015). *A qualitative comparative study of Norwegian and Hawaiian knowledge workers* [Unpublished Master's thesis]. Trondheim: NTNU.

Frese, M., & Fay, D. (2001). Personal initiative: An active performance concept for work in the 21st century. *Research in Organizational Behavior, 23,* 133–187.

Frese, M., Fay, D., Hilburger, K. L., & Tag, A. (1997). The concept of personal initiative: Operationalization, reliability and validity in two German samples. *Journal of Occupational and Organizational Psychology, 70,* 139–161.

Frese, M., Kring, W., Soose, A., & Zempel, J. (1996). Personal initiative at work: Differences between East and West Germany. *The Academy of Management Journal, 39*(1), 37–63.

Frese, M., Teng, E., & Wijnen, C. J. D. (1999). Helping to improve suggestion systems: Predictors of making suggestions in companies. *Journal of Organizational Behavior, 20,* 1139–1155.

Frese, M., & Zapf, D. (1994). Action as the core of work psychology: A German approach. In H. C. Triandis, M. D. Dunnette, & L. Hough (Eds.), *Handbook of industrial and organizational psychology* (Vol. 4, pp. 271–340). Palo Alto: Consulting Psychologists Press.

George, J. M. (1996). Group affective tone. In M. A. West (Ed.), *Handbook of work group psychology* (pp. 77–94). New York: Wiley.

Goodman, S. A., & Svyantek, D. J. (1999). Person-organization fit and contextual performance: Do shared values matter? *Journal of Vocational Behavior, 55,* 254–275.

Gouldner, A. W. (1960). The norm of reciprocity: A preliminary statement. *American Sociological Review, 25,* 161–179.

Hollander, E. P., & Webb, W. E. (1955). Leadership, followership, and friendship: An analysis of peer nominations. *Journal of Abnormal and Social Psychology, 50*(2), 163–167.

Katz, D. (1964). The motivational basis of organizational behavior. *Behavioral Science, 9,* 131–146.

Kelley, R. (1992). *The power of followership.* New York: Doubleday.

Kelley, R. E. (1988). In praise of followers. *Harvard Business Review, 66,* 142–148.

Kickul, J., & Gundry, L. K. (2002). Prospecting for strategic advantage: The proactive entrepeneurial personality and small firm innovation. *Journal of Small Business Management, 40,* 85–97.

Kilhammar, K., & Ellström, E. (2015). Co-workership in practice: A study of two Swedish organizations. *Human Resource Development International, 18*(4), 328–345.

LePine, J. A., Erez, A., & Johnson, D. E. (2002). The nature and dimensionality of organizational citizenship behavior: A critical review and meta-analysis. *Journal of Applied Psychology, 87* (1), 52–65.

Morrison, E. W., & Phelps, C. C. (1999). Taking charge at work: Extra-role efforts to initiate workplace change. *Academy of Management Journal, 42,* 403–419.

Organ, D. W. (1988). *Organizational citizenship behavior: The good soldier syndrome.* Lexington: Lexington Books.

Organ, D. W. (1990). The motivational basis of organizational citizenship behavior. *Research in Organizational Behavior, 12*(1), 43–72.

Organ, D. W. (1997). Organizational citizenship behavior: It's construct clean-up time. *Human Performance, 10*(2), 85–97.

Parker, S. K., William, H. M., & Turner, N. (2006). Modeling the antecedents of proactive behavior at work. *Journal of Applied Psychology, 91*(3), 636–652.

Podsakoff, N. P., MacKenzie, S. B., Moorman, R. H., & Fetter, R. (1990). Transformational leader behaviors and their effects on followers' trust in leader, satisfaction, and organizational citizenship behaviors. *Leadership Quarterly, 1*(2), 107–142.

Podsakoff, N. P., Whiting, S. W., Podsakoff, P. M., & Blume, B. D. (2009). Individual and organizational-level consequences of organizational citizenship behaviors: A meta-analysis. *Journal of Applied Psychology, 94*(1), 122–141.

Robinson, S. L. (1996). Trust and breach of the psychological contract. *Administrative Science Quarterly, 41*(4), 574–599.

Rousseau, D. M. (1989). Psychological contracts in organizations. *Employee Responsibilities and Rights Journal, 2*(2), 121–139.

Simard, M., & Marchand, A. (1995). A multilevel analysis of organizational factors related to the taking of safety initiatives by work groups. *Safety Science, 21,* 113–129.

Skogstad, A. (2005). Den psykologiske kontrakt mellom arbeidstaker og arbeidsgiver [The psychological contract between employee and employer]. In S. Einarsen & A. Skogstad (Eds.), *Den dyktige medarbeider* [The competent employee], (pp. 273–296). Bergen: Fagbokforlaget.

Somech, A., & Drach-Zahavy, A. (2000). Understanding extra-role behavior in schools: The relationship between job satisfaction, sense of efficacy, and teachers' extra-role behavior. *Teaching and Teacher Education, 16,* 649–659.

Tengblad, S. (2010). *Medarbetarskap på 60 minuter.* Retrieved from http://www.diva-portal.org/smash/get/diva2:619613/FULLTEXT01.pdf.

Van Dyne, L., Graham, J. W., & Dienesch, R. M. (1994). Organizational citizenship behavior: Construct redefinition, operationalization, and validation. *Academy of Management Journal, 37,* 765–802.

Van Dyne, L., & LePine, J. A. (1998). Helping and voice extra-role behaviors: Evidence of construct and predictive validity. *Academy of Management Journal, 41*(1), 108–119.

Chapter 10
Successful Aging at Work

Karianne Kvalheim, Marit Christensen and Siw Tone Innstrand

Life expectancy is increasing all over the globe; in Norway, a person born in 2013 can expect to live into his or her 80s (Statistics Norway, 2014). As the population expectations for a longer life accumulate, it has been found that people are taking lengthened age époques into consideration when planning their lives (WHO, 2015). We have already seen an increase in higher education (Statistics Norway, 2016), which results in a later entry into working life. At the same time, the population is aging; in Norway, the estimate of the proportion of people older than 70 years will increase from about 11% today to 19% in 2060 (Tønnessen, Syse, & Aase, 2014). A growing concern is the lack of active workers to sustain the welfare system, as the outlook shows that in 2050, assuming moderate growth, there will be in excess of two active workers per pensioner (Brunborg, 2004; Østby, 2004). In contrast, in 1950, were there were more than seven people of working age per pensioner (NIPH, 2008). At a time when more people are leaving the work force than entering it, understanding how to enable and motivate seniors to remain engaged, healthy, and productive in work will be of major importance.

It is a known fact that cognition, physical capabilities, and mental health are likely to decline with advancing age. Still, several studies have found that this deterioration is not homogeneous across all individuals as some individuals maintain cognitive capabilities and physical and mental health even into their later

K. Kvalheim (✉) · M. Christensen
Department of Psychology, Norwegian University of Science and Technology (NTNU), Trondheim, Norway
e-mail: karianne.kvalheim@ntnu.no

M. Christensen
e-mail: marit.christensen@ntnu.no

S.T. Innstrand
Department of Public Health and Nursing, Faculty of Medicineand Health Sciences, Norwegian University of Science and Technology, Trondheim, Norway
e-mail: siw.tone.innstrand@ntnu.no

M. Christensen et al. (eds.), *The Positive Side of Occupational Health Psychology*,
https://doi.org/10.1007/978-3-319-66781-2_10

years (Wickrama, O'Neal, Kwag, & Lee, 2013), and also that working in later life may have a positive influence on the physical functioning of older adults. For example, whilst controlling for socio-economic characteristics, Hinterlong, Morrow-Howell, and Rozario (2007) found that being a paid worker or working as a volunteer was associated with better self-rated health and less functional impairment in adults older than 60 years. Despite the beneficial aspects of staying healthy and productive in work, in general, and the need to maintain a healthy workforce in an aging population, there is a lack of knowledge about successful aging at work.

In this chapter, we will explore the concept of successful aging at work (SAW). We will suggest a working definition and explain some of the challenges of studying SAW and how research on this concept could illuminate possible solutions to the challenges faced by society, the organizations, and the individual in order to succeed with successful aging at work. Finally, we will provide some suggestions on how to develop best practice and what we believe will be important for further research.

10.1 The Concept of Successful Aging at Work

A precondition for remaining in work in increasing older age is, of course, the employees' working abilities, and a large part of this pertains to health. In this regard, it is appropriate to take into account that it is more likely that disease and health complaints occur, and also endure, as people get older. Still, only 16% of seniors consider their health as poor or very poor, despite the fact that 73% of seniors have one or more permanent illnesses (Ugreninov, 2005). There is also evidence that having a stronger sense of positive well-being may be a resource for healthier aging: slower physical decline, reduced risk of frailty, and longer survival (Allerhand, Gale, & Deary, 2014). This indicates that successful and healthy aging is more than just the absence of disease, and leads us to our working definition of successful aging at work. By successful aging at work, we mean a workplace where seniors can utilize and develop their resources optimally (e.g., health, competence), feel that they master different demands, and experience meaning and well-being at work. As well as being engaged in the job, they want to actively participate and they delay withdrawal from the labor market. From the organization's perspective, successful aging constitutes a full utilization of the working capacity where knowledge retention and transfer are also key elements. From a societal perspective, a full utilization of the society's work resources, from a health perspective, is of major importance. Successful aging at work is by this definition an important prerequisite for a sustainable welfare society.

Some of the challenges of studying successful aging at work have been the lack of consensus on the operational definition, but also on the questions of which elements are central and what constitutes successful aging. Accordingly, different outcome measures are used. There are also different opinions at to which specific criteria should be used to evaluate successful aging: subjective and/or objective, age-related explanatory mechanisms, facilitating and constraining factors/conditions, or temporal patterns (Zacher, 2015). Furthermore, there is no consensus in the literature about who is and who is not an older worker. The definition varies across historical periods and industrial sectors. For example, in the IT sector (in the US) workers are considered old if they have children. Professional athletes may be considered to be old in their 20s or 30s, whereas airline pilots and firemen are considered old in their 50s, and Supreme Court Justices (in the US) in their 80s. The meaning of aging has also changed as a substantial percentage of today's seniors are healthier and active. Seniors in their 60s, 70s, and 80s have increased vitality. This has altered the responsibilities, ambitions, daily activities, and lifestyles of these groups (Pitt-Catsouphes, & Smyer, 2006) and might also alter our views as to whom we refer to as elderly. A survey on American Perceptions of Aging revealed that men were considered to be "old" at a median age of 70 years, compared to 75 years for women (Cutler & Whitelaw, 2002). To avoid dealing with this inconsistent language label on older age, some authors present their research results by age groups, such as 25–39 year olds, 40–54 year olds, and 55 plus (Pitt-Catsouphes, & Smyer, 2006), or under 30, 30–49, and over 50 (e.g., in Anthun & Innstrand, 2015), and some statistical analytic organizations often use 5 or 10 year intervals when presenting numbers (e.g., Statistics Norway). It might be important for the employers and budget planners to remember that a prolonged education, where increasing numbers of people tend to gain higher degrees at university level, results in a later entry into the labor market. In earlier decades, mid-career workers seemed to become older workers once they began planning for retirement. At age 50, the employee might be in the middle of his or her career and not even close to the end of working life. Still, it has been suggested that some workers already in their 50s have started planning for retirement, and new research has found that older workers tend to have different needs than their younger colleagues (see e.g., Anthun & Innstrand, 2015). To understand and to create a healthy work environment, knowledge of these differences and what they imply will be of importance for practitioners and further research.

In order to illuminate the age dimension on successful aging, Zacher (2015) has provided a theoretical framework of how the interplay between age, the age-related mediating mechanisms, and the facilitating and constraining factors influence successful aging over time (see Fig. 10.1). According to Zacher's model, "age is associated with person characteristics (e.g., knowledge, skills, abilities, and other factors such as personality, motivation, and interests) as well as work characteristics (e.g., task, job, team, organization, vocation factors) and life circumstances (e.g., family, hobbies, volunteering activities)" (Zacher, 2015, p. 12). The dashed arrows in the figure indicate that the employee's age is not a causal variable but represents the intra-individual age-related changes over time.

This model illustrates some of the complexity but also illuminates the mediating and moderating elements that we need to consider in order to influence work outcomes in intended direction. The aging of the population generates new elements and questions, and we might need new and possibly divergent solutions.

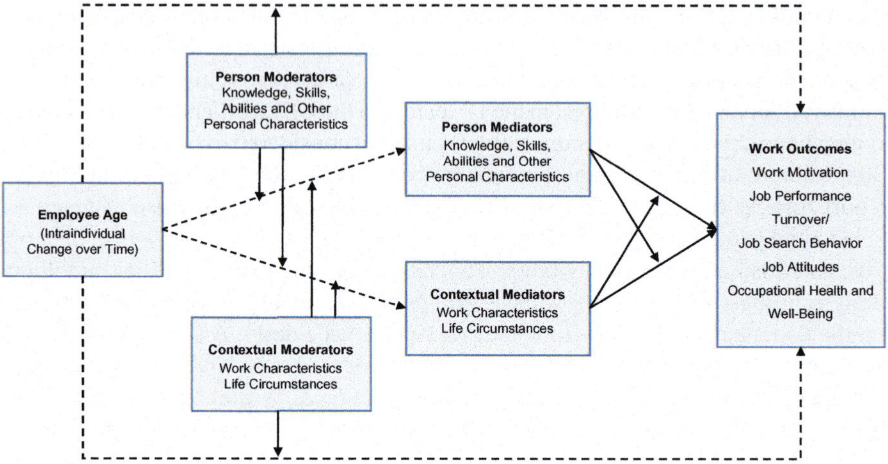

Fig. 10.1 "Zacher's theoretical framework of successful aging at work." *Source* Zacher (2015)

10.2 Challenges in Society Today

The World Health Organization recommends a societal approach to population aging, which includes the goal of building an age-friendly world. They also emphasize the importance of a transformation away from disease-based curative models and toward the provision of the needs of (older) people (WHO, 2015). The WHO report is mainly targeted toward health care systems, but is also relevant for organizations and older workers in society as it integrates the health definition from 1946 that defines health as a "state of complete physical, mental, and social well-being and not merely the absence of disease or infirmity" (WHO, 1946, p. 100). This shift broadens how we (society, organizations, and individuals) ought to see health at work, and at the same time, it captures needs of the individual. The theoretical and empirical understanding of the concept of health is of great importance for best practice and underlines the need for a better linking of research and practice in this field. As the aging of populations is rapidly accelerating worldwide (WHO, 2015), the cost to society and the challenges that follow raise questions. For example, in Norway, the estimated cost to society of one person retiring early is 134% of the person's previous salary before tax (Holmøy, 2002). We have also seen an increasing number of people exiting the labor market and entering into long-term social security schemes, and in Norway alone, the government spends approximately NOK 30 billion each year on sickness and benefit payments. Still, according to WHO (2015) "the greatest costs to society are not the expenditures made to foster (this) functional ability, but the benefits that might be missed if we fail to make the appropriate adaptations and investments" (WHO, 2015 p. 4).

Work life might be a great source of social and intellectual stimulation, by both involving scheduled constrains and requiring the allocation of cognitive recourses, which in turn may support brain function, also into older adulthood. There are also findings that an increasing retirement age trends toward a reduced risk of dementia, but some of this research has been criticized for selection bias, and we need more knowledge to conclude (Grotz, Meillon, Amieva, & Letenneur, 2015). However scarce the evidence might be at this stage, the suggestion that work itself, and how it is organized, may influence not only how long an employee is able to work but also the health trajectory for an employee, is very interesting. With this in mind, we might say that society's goals, in terms of economy, for a healthy population are coinciding with the older workers needs and health.

10.3 How to Succeed: A Way to Best Practice

According to the Norwegian Working Environment Act, the working environment should provide "a basis for a healthy and meaningful working situation, that affords full safety from harmful physical and mental influences" (…) and "facilitate adaptations of the individual employee's working situation in relation to his or her capabilities and circumstances of life" (Work Act, 2005, § 1). This means that the organizations in Norway are legally required to assist the individual. In so doing, it will be advantageous to ensure that the intention of the assistance and the result correspond. To exemplify, one common type of action that is intended to lower absenteeism is to reduce the demands on employees in the workplace. Nevertheless, the intention and the result are not necessarily equal, as several studies have shown a tendency that a reduction in job demands alone does not necessarily lead to higher job satisfaction or engagement. Accordingly, reducing demands alone might not lead to increased job presence.

If we look at the research, we see that several studies found that job resources are important, not only to reduce the negative impact of high job demands, but they are also important in their own right. According to the Job Demands-Resources (JD-R) model, job resources appear to drive a motivating process that leads to job-related learning, work engagement, and organizational commitment (see, for example Demerouti, Bakker, de Jonge, Janssen, & Schaufeli, 2001; Salanova, Agut, & Peiró, 2005; Taris & Feij, 2004). The JD-R model (see, e.g., Chaps. 3 and 7 of this book) is based on the assumption that whereas every occupation may have its own specific risk factors associated with job stress, these factors can be classified in two general categories (job demands and job resources), thus constituting an overarching model that may be applied to various occupational settings irrespective of the particular demands and resources involved. It is well established that work engagement is an important predictor for positive health, for job performance overtime, and also that job resources predict work engagement (Avery, McKay, & Wilson, 2007; Demerouti et al., 2001; Hakanen, 2009; Salanova et al., 2005; Schaufeli & Bakker, 2004; see also Chap. 7 of this book).

Age and gender differences in engagement have not been the main focus of research but a study by Haley, Mostert, and Els (2013) found that older workers experienced lower levels of exhaustion compared to their younger colleagues, and that it was different job demands and resources that predicted engagement dependent on age (groups) in the South African financial sector. Anthun and Innstrand (2015) also found that the motivation to work was determined by different factors dependent on age groups. For example, recognition and autonomy at work were more important for older employee's motivation than it was for their younger colleagues.

Such research-based knowledge is crucial in order to target the different varieties of facilitation and interventions. And in this light, there is a paradox in that many Norwegian businesses offer the same type of action, regardless of job type or task (Midtsundstad & Nielsen, 2014), whereas the research indicates that dissimilar actions should be taken to strengthen the work motivation of different age groups. This kind of knowledge is a key element when trying to create more engaged and healthy (older) workers in terms of knowing which factors needs to be focused on and when. Thus, it is of great important for organizations to base their practice of successful aging on research-based knowledge. Further, organizations should also play a role in the process of generating new knowledge about what works and what does not work through thorough process evaluation and effect evaluation of their interventions. Not surprisingly, there is evidence to support the suggestion that work ability—the capacity to meet the physical, mental, and social demands of a job—is linked to positive health outcomes for older workers (Kooij, 2015). Interestingly, Bohle, Pitts, and Quinlan (2010) found that work characteristics are more critical than worker's individual capacities to maintain positive health at work. This implies that work and how it is organized may have an effect on the employee's health and probably on the length of time that an employee is able and motivated to work.

10.4 Successful Aging and the Future

To understand and develop a best practice for successful aging at work in the future, close attention needs to be paid to the results of research in this area. A best practice for successful aging is a workplace where (older) workers can utilize and develop their resources optimally, experience mastery, meaning, and well-being at work. So far, we know that different age groups need different measures in the workplace in order to maintain positive health, well-being, and productivity in the workplace. Older workers are also different depending upon, for example, the occupation, work hours, and personal factors. In order to facilitate successful aging at work, it will be important to investigate and explain individual and organizational factors that facilitate and constrain well-being, health, and labor participation, especially among older workers.

Furthermore, motivation is a significant factor to facilitate successful aging at work; this is important not only for the employees but is also an important tool for the organization, where successful aging constitutes a full utilization of the working capacity in which knowledge retention and transfer are key elements. Research has shown that reducing demands is not enough and attention also needs to be paid to job resources in order to develop a healthy and productive work environment for older workers. Further research needs to investigate which resources are the most important for older workers in different occupations and organizations in order to maintain successful aging at work. From a societal perspective, a full utilization of the society's work resources, from a health perspective, is of major importance. A Norwegian report entitled "Effects of bridge employment on health" and written for the Ministry of Health and Care Services, concludes that some individuals will benefit from early retirement, whereas others will gain positive health effects from having the opportunity to continue to work beyond retirement age. Additionally, this report pointed to the fact that workers and occupations differ and that there might be differences between occupations—an important factor that requires further investigation (Smedslund, Steiro, & Forsetlund, 2014) in order to generate more targeted actions to achieve successful aging at work. Successful aging at work is, by this definition, an important prerequisite for a sustainable welfare society.

Accordingly, we need to know more about the differences between occupations and sectors. It could be interesting to explore what characterizes the sectors and organizations that succeed with aging in the workplace and whether there are elements that other organizations and sectors can utilize. Given the importance of cohort influences such as educational achievement, improved health and physical function, and technological advance, longitudinal studies might provide a different picture. It will thus also be important to identify which resources are most important for the older workers to strengthen the motivation process (cf. JD-R model). Some crucial research questions are: what are the determinants for inclusion in work life among older employees? Does the predicted value of job demands and resources on health and engagement differ across age groups? And what characterizes the workers, organizations, and sectors that succeed with aging at work overtime?

10.5 Final Remarks

In general, employees are aging successfully at work if "they are able to preserve and regenerate resources to achieve their personal goals maintain their health, motivation, and work ability now and in the future" (Kooij, 2015, p. 310). The potential for work to be beneficial for a person's health is intriguing. Knowledge about the individual and organizational factors that facilitate and constrain well-being, health, and labor participation among older workers is of major significance for the government, with regard to planning budgets and also legislation in the longer run, for the organizations, with regard to loss of resources, productivity, and profits, and, of course, for the employees regarding their health, meaning, inclusion, and quality of life. Research-based knowledge on the conditions for health promoting work places and successful aging will contribute to long-term human resource development and to developing the knowledge base for policy in this area.

References

Allerhand, M., Gale, C. R., & Deary, I. J. (2014). The Dynamic Relationship between cognitive function and positive well-being in older people: A prospective study using the English longitudinal study of aging. *Psychology and Aging, 29*(2), 306–318. doi:10.1037/a0036551.

Anthun, K. S., & Innstrand, S. T. (2015). The predictive value of job demands and resources on meaning of work and organizational commitment across different age groups in the higher education sector. *Journal of Higher Education Policy and Management, 38 (1).*

Avery, D. R., McKay, P. F., & Wilson, D. C. (2007). Engaging the aging workforce: The relationship between perceived age similarity, satisfaction with coworkers, and employee engagement. *Journal of Applied Psychology, 92*(6), 1542–1556.

Bohle, P., Pitts, C., & Quinlan, M. (2010). Time to call it quits? The safety and health of older workers. *International Journal of Health Services, 40*(1), 23–41.

Brunborg, H. (2004). *Befolkningen blir eldre* [The population is aging]. Økonomiske analyser 5/2004, Statistisk sentralbyrå.

Cutler, N. E., and Whitelaw, N. A. (2002). American Perceptions of Aging in the 21st Century. Ann Arbor: Inter-university Consortium for Political and Social Research. doi:http://doi.org/10.3886/ICPSR03326.v1.

Demerouti, E., Bakker, A. B., De Jonge, J., Janssen, P. P. M., & Schaufeli, W. B. (2001). Burnout and engagement at work as a function of demands and control. *Scandinavian Journal of Work, Environment & Health, 27,* 279–286.

Grotz, C., Meillon, C., Amieva, H., & Letenneur, L. (2015). Why is later age at retirement beneficial for cognition? Results from a French population-based study. *The Journal of Nutrition Health and Aging, 20*(5). doi:10.1007/s12603-015-0599-4.

Hakanen, J. J. (2009). Do engaged employees perform better at work? The motivating power of job resources and work engagement on future job performance. In M. Christensen (Ed.), *Validation and test of central concepts in positive work an organizational psychology—The second report of the Nordic Project.* (TemaNord 2009:564, pp. 65–71). København: Nordic Council of Ministers.

Haley, L.-M., Mostert, K., & Els, C. (2013). Burnout and work engagement for different age groups: Examining group-level differences and predictors. *Journal of Psychology in Africa, 23* (2), 283–295. doi:10.1080/14330237.2013.10820625.

Hinterlong, J. E., Morrow-Howell, N., & Rozario, P. A. (2007). Productive engagement and late life physical and mental health. findings from a nationally representative panel study. *Research on Aging, 29*(4), 348–370.

Holmøy, E. (2002). *Hva koster tidligpensjonering for samfunnet?* [What does early retirement cost for society?]. Økonomiske analyser 2/2002, Statistic Norway.

Kooij, D. T. A. M. (2015). Successful aging at work: The active role of employees. *Work, Aging and Retirement, 1*(4). doi:10.1093/workar/wav018.

Midtsundstad, T., & Nielsen, R. A. (2014). Arbeid og pensjon i kommunal sektor. Analyse av survey- og registerdata fra før og etter pensjonsreformen [Work and Pensions in the municipal sector. Analysis of survey and register data from before and after the pension reform]. *Fafo-rapport 2014:45.*

NIPH. (2008). *Levealder i Norge, fakta om forventet levealder* [Life expectancy in Norway, facts about life expectancy]. Downloaded 06.11.2009 from: http://www.fhi.no/eway/default.aspx?pid=233&trg=MainLeft_5648&MainArea_5661=5648:0:15,2917:1:0:0:::0:0&MainLeft_5648=5544:72449::1:5647:43:::0:0.

Østby, L. (2004). *Ikke blant Europas største, men dyrt kan det bli* [Not among Europe's largest, but expensive it might get]. Samfunnsspeilet 1, 2004, Statistics Norway.

Pitt-Catsouphes, M., & Smyer, M. A. (2006). *How old are today's older workers?* (Issue Brief No. 4). Chestnut Hill, MA: The Center on Aging and Work/Workplace Flexibility at Boston College.

Salanova, M., Agut, S., & Peiró, J. M. (2005). Linking organizational resources and Work engagement to employee performance and customer loyalty: The mediating role of service climate. *Journal of Applied Psychology, 90,* 1217–1227.

Schaufeli, W. B., & Bakker, A. B. (2004). Job demands, job resources, and their relationship with burnout and engagement: A multi-sample study. *Journal of Organizational Behavior, 25,* 293–315.

Smedslund, G., Steiro, A. K., & Forsetlund, L. (2014). *Helseeffekter av å fortsette i arbeid etter pensjonsalder* [Health effects of continuing to work after retirement]. Notat. ISBN 978-82-8121-850-5.

Statistics Norway, (2014). *Forventet levealder ved fødselen. Indikatorer for bærekraftig utvikling, 2014* [Life expectancy at birth. Indicators for Sustainable Development, 2014]. Downloaded 10.07.2016 from: http://www.ssb.no/natur-og-miljo/barekraft/forventet-levealder-ved-fodselen.

Statistics Norway. (2016). *Population's level of education, 1 October 2015*. Downloaded 17.08.2016 from: https://www.ssb.no/en/utdanning/statistikker/utniv/aar/2016-06-20.

Taris, T. W., & Feij, J. A. (2004). Learning and strain in newcomers: A three-wave study on the effects of job demands and job control. *Journal of Psychology, 138,* 543–563.

Tønnessen, M., Syse, A., & Aase, K. N. (2014). Befolkningsframskrivinger 2014–2100: Hovedresultater. *Økonomiske analyser 4/2014.*

Ugreninov, E. (2005). *Seniorer og arbeid. Dessertgenerasjonen; lever lenger og jobber kortere* [Seniors and work. The Dessert generation; living longer and working less]. Samfunnsspeilet 5, 2005, Statistics Norway.

WHO. (1946). *Official records of the World Health Organization*. Preamble to the Constitution of the World Health Organization as adopted by the International Health Conference, New York.

WHO. (2015). *World report on ageing and health.* ISBN: 978 92 4 156504 2. http://apps.who.int/iris/bitstream/10665/186463/1/9789240694811_eng.pdf.

Wickrama, K. K., O'Neal, C. W., Kwag, K. H., & Lee, T. K. (2013). Is working later in life good or bad for health? An investigation of multiple health outcomes. *Journal of Gerontology: Psychological Sciences, 68*(5), 807–815. doi:10.1093/geronb/gbt069.

Work Act. (2005). Lov, 17. juni 2005 nr. 609, om arbeidsmiljø, arbeidstid og stillingsvern mv [Act relating to working environment, working hours and employment protection, etc. (Working Environment Act)]. Oslo: Arbeidsdepartementet.

Zacher, H. (2015). Successful aging at work. *Work, Aging and Retirement, 1*(1), 4–25. doi:10.1093/workar/wau006.

Chapter 11
From Sickness Absenteeism to Presenteeism

Per Øystein Saksvik, Karoline Grødal and Maria Karanika-Murray

Throughout the past decades, sickness absenteeism has received increased attention in Norway. An important motive for this has been that sickness absenteeism is relatively easily converted into costs. There is no doubt that the costs are huge and are clearly visible in the national budget. The Ministry of Labor in Norway suggested that a grant of NOK 38,520,720,000 (4,147,275,037 €) be used for sick pay in the national budget for 2014 (representing around 1.9% of the national budget). This sum is not even the full cost of sickness absenteeism, but rather it represents the part that the public system refunds or pays for sickness absenteeism throughout the working period (costs after 16 days of continued sick leave). This is a great share of the absenteeism, around 80%, and is a number we calculated before the new millennium (Lund, Rognstad, Rundmo, & Saksvik, 1991). Still, since then, there have not been any noteworthy changes in the sickness absenteeism system in Norway, and so, the share is probably still the same today. Therefore, the state pays the largest part of the sickness absenteeism cost, and it is discussed on a regular basis whether the employer (paying the remaining 20%) and even the employee should contribute more to the costs.

P.Ø. Saksvik (✉)
Department of Psychology, Norwegian University of Science and Technology (NTNU), Trondheim, Norway
e-mail: per.saksvik@ntnu.no

K. Grødal
Department of Public Health and Nursing, Faculty of Medicine and Health Sciences, Norwegian University of Science and Technology, PO Box 8905, Trondheim 7491, Norway
e-mail: karoline.grodal@ntnu.no

M. Karanika-Murray
Department of Psychology, Nottingham Trent University, Nottingham, UK
e-mail: maria.karanika-murray@ntu.ac.uk

M. Christensen et al. (eds.), *The Positive Side of Occupational Health Psychology*,
https://doi.org/10.1007/978-3-319-66781-2_11

11.1 Sickness Absenteeism and Presenteeism

The term presenteeism has been the subject of growing interest in the field and is largely related to sickness absenteeism. This chapter concerns presenteeism and what this term contributes to the debate about sickness absenteeism. A commonly accepted definition of presenteeism states that the job should be arranged so that you can manage the tasks while sick (or undertake another job that is possible to manage while sick) without worsening the condition, or preferably, while improving the condition. Presenteeism has traditionally in the literature been linked to a decrease in productivity/ability to work (Biron & Saksvik, 2010). The reason for this is that the difficulties in pressuring oneself to work when one is not feeling well can lead to not performing as effectively as possible and to a risk of increased sickness absenteeism in the future (Lindberg, Josephson, Alfredson, & Vingård, 2008; Thun, Saksvik, Ose, Mehmetogu, & Christensen, 2013).

11.2 The Authorities' Initiative for Reducing Sickness Leave

In the international research, a negative assumption concerning presenteeism is dominant. The prevailing view is that presenteeism can only be negative for productivity. Therefore, it is paradoxical that Norwegian authorities wish to stimulate a more extensive use of presenteeism by establishing arrangements such as graded sick leave. These kinds of arrangements presuppose that a person that normally would be on 100% sick leave now works a reduced day which requires presenteeism. As long as the established sickness benefit arrangement is not to be affected, it is understandable that the authorities would undertake these measures. Now, the authorities need to understand under which circumstances presenteeism can be something positive. Our own research shows how good facilitation can make presenteeism positive (Rostad, Milch, & Saksvik, 2015). Karanika-Murray & Biron have undertaken a critical review of presenteeism where the fact that presenteeism can be both deleterious, but also functional for both health and performance is highlighted and present a model for that Karanika-Murray and Biron (2016). This chapter will deal with the relationship between presenteeism and sickness absenteeism and the underlying mechanism for presenteeism to operate on a positive level.

The Norwegian authorities' initiative for limiting sickness absenteeism has a clear link to the state paying for most of the sickness absenteeism. Through the years, it has been suggested that Norway should establish "Karen's days," as was done in Sweden in 1993, meaning that some of the costs of sickness absenteeism are paid by the employee. This means that the state does not pay for the first day of the sickness absenteeism (as it does now). After the initiation of "Karen's days," sickness absenteeism in Sweden decreased, but today, we find about the same level as in Norway (Lie, 2009). In our part of the world, Sweden, Norway, and the Netherlands have the highest number of sick leave days (in relation to the EU countries), which is purported to have arisen because of the countries' favorable sick leave benefits (Lie, 2009). The new conservative government in Norway has stated that the sick leave benefits are not going to change and that the authorities have to find different initiatives to reduce the costs. Graded sick leave is an initiative that has been proposed to reduce these costs. Norwegian labor welfare administrations recommend graded sick leave and conducted their own research that showed a 2% drop in sick leaves (Bjørgum, Kristoffersen, & Kalajdzic, 2012), a drop in sick leave that is not substantial, but may be a contribution in the right direction. The initiative is debated, mostly because of the many conditions for it to work properly. One condition concerns adjusting the working conditions, so that the employee does a good job which does not lead to further worsening of their health condition. Implicitly, this indicates that the workload is approximately normal when the employee is at work, so that the presenteeism is not a burden for colleagues or that the employer incurs additional costs. This situation makes the need for a discussion about presenteeism even more important. Markussen, Mykletun, and Røed (2012) found that workers on graded sick leave, who were able to use their remaining employability because of facilitation and an ability to work, had a shorter period of sickness absenteeism in total. In the same way, facilitation is also important for those who are on long-term sickness absenteeism and want to return to work, which thus affects the decision to engage in presenteeism or absenteeism. The problems related to adjusting work to fit the health situation of the presentee may be a complex and challenging one both for the employer and the colleagues.

11.3 Presenteeism and Productivity

Most of the research done on presenteeism has a negative orientation toward presenteeism as a problem for productivity. The reason for this view is that the employees go to work in a condition of being less effective than normal. The research highlights, for example, the direct loss in productivity (Cooper & Dewe, 2008), the organization of working hours (overtime increases the chances of presenteeism) (Bökerman & Laukkanen, 2010a), individual decisions about taking a sick leave or staying on the job while sick (Wynne-Jones, Buck, Varanava, Phillips, & Main, 2009), the type of health issue (Aronsson, Gustafsson, & Dallner, 2000), and the profession (Elstad & Vabø, 2008; McKevitt, Morgan, Dundas, & Holland, 1997). The literature gives a one-sided picture of presenteeism as something that should be prevented, at any cost; it is a risk factor for any business and the individual employee. The main reason for this perspective may be that adjusting the work situation for the presentee seldom is considered. This may come as a consequence of a cost-effective estimate. Since the state in Norway pays almost all costs related to absenteeism, the authorities must leave no stone unturned to find optimal solutions. This means that adjusting work has to be considered to reduce costs indicating that the state has to stimulate to this activity at the business level by introducing measures like graded sick leave, the IWL agreement (see below), and other initiatives.

11.4 Presenteeism and Attendance Pressure

Several studies have tried to investigate the course of presenteeism in a more negative light insofar as it is a condition that forces one to stay on the job when an employee, because of health reasons, should remain at home (Aronsson & Gustafsson, 2005; Hansen & Andersen, 2008; Johns, 2011). Saksvik (1996) conducted a study on attendance pressure in an organization going through restructuring and identified four pressure factors: importance pressure, sanction pressure, moral pressure, and safety pressure. Even though these factors were found in a special context—a restructuring business—it is possible that these factors could be more or less universal since there is a large amount of job change and an uncertainty in today's working life. Grødal (2014) has examined these factors more closely and summarized them in relation to new literature. Importance pressure occurs when one has a lot to do, is difficult to be replaced, or has an important task in the job (Saksvik, 1996). Several studies have shown that time pressure and a large workload have a clear relationship to presenteeism (Aronsson & Gustafsson, 2005; Biron, Brun, Ivers, & Cooper, 2006; Claes, 2011; Hansen & Andersen, 2008). In addition, it seems as if a large level of responsibility in the job and the notion of the work meaning a lot to others (e.g., to colleagues, customers, clients, or patients) can contribute to a greater frequency of presenteeism (Gosselin, Lemyre, & Corneil, 2013; Johns, 2011). Some employees choose presenteeism, because they think sickness absenteeism can result in negative consequences for themselves or others for whom they have responsibility (Grinyer & Singleton, 2000). If one does not have the opportunity to be replaced when absent, one can experience a greater workload when returning to the job (Aronsson & Gustafsson, 2005). This seems to have a connection with higher presenteeism (Aronsson, Gustafsson, & Dallner, 2000; Aronsson & Gustafsson, 2005; Biron et al., 2006, Bökerman & Laukkanen, 2010b). A study done by Johns (2011), however, found the opposite—that greater presenteeism was related to the feeling of being replaceable when absent. Results from the same study also showed that mutual dependence among members of a work team could have a connection with cases of presenteeism, but no connection with the length of presenteeism.

Sanction pressure occurs when one fears being accused of an illegal absence by the management or colleagues as, for example, when staying home from the job without a reason (Saksvik, 1996). Milch (2011) found that this factor was a significant predictor of presenteeism. In a quantitative study of female leaders, researchers found that several of the respondents sacrificed their spare time even though there was not a big workload, since they thought it was expected of them

(Simpson, 1998). One respondent, for example, said that she felt her colleagues would notice if she left work before them and that they would think, “there she goes again!” (p. 44). Baker-McClearn, Greasley, Dale, and Griffith (2010) highlight that employees can feel the pressure of presenteeism if they risk financial losses by not showing up at work. Additionally, the organization can create a culture that promotes unfavorable presenteeism because of the organization’s practice of sick leave and salary schemes. Another example is that the follow-up talks can become pressing for the sick individual and contribute to them returning to work earlier than they should. The same applies if the management belittles the employee’s health condition. Grinyer and Singleton (2000) found that systems where employees are called in after having a set number of sick leaves, could feel like a punishment, and cause a fear of taking a sick leave. In a study by Biron et al. (2006), fear of a negative outcome was one of the reasons stated for choosing to go to work even while sick.

Moral pressure covers upbringing and attitude, when the employee has a sense of responsibility and has a bad conscience when they do not show up (Saksvik, 1996). In a study by Biron et al. (2006), guilt was one of the most frequently stated reasons for choosing presenteeism. Baker-McClearn et al. (2010) conducted a qualitative study in which they interviewed individuals with different professions in Great Britain. They found that the sense of commitment and loyalty toward the organization and the team was important when choosing to come to work when sick. Many had a feeling of letting down their colleagues if they stayed home from work. Some even felt that an absence was an indicator of bad accomplishments and feared that this could affect their chances of promotion. Quantitative studies have found that accomplishment-based confidence has a positive correlation with presenteeism (Aronsson & Gustafsson, 2005; Löve, Grimby-Ekman, Eklöf, Hagberg, & Develle, 2010). Similarly, Collins and Cartwright (2012) found that moral pressure and having a conscience toward coworkers could contribute to presenteeism, but they also found that motivation for the job had an impact. Some were concerned with finishing a project, and some had difficulty with delegating work to someone else when they had put their soul into it and had spent a long time on a task.

Security pressure occurs when one is uncertain and perceives a fear of losing a job if they stay away (Saksvik, 1996). Milch (2011) found that this factor was significantly correlated with, but not a predictor of, presenteeism. Studies have shown that job uncertainty has a relationship with presenteeism (Hansen & Andersen, 2009; Johns, 2011). Heponiemi et al. (2010), however, found that the employee’s experience of a situation plays a greater part than the actual level of job uncertainty when it comes to the propensity for presenteeism.

11.5 Absence Legitimacy

Absence legitimacy will also be presented as a factor that could have an impact on the choice of sickness absenteeism or presenteeism. Absence legitimacy is bound to the social context and refers to which level the employee perceives the absence from work as favorable or unfavorable behavior (Addae & Johns, 2002; Dahl-Jörgensen, Moe, & Saksvik, 2002). Johns (2010) postulated that absence legitimacy is an individual factor that can influence the choice between absenteeism and presenteeism. Later, he found that absence legitimacy had a positive correlation with the frequency of presenteeism (Johns, 2011). Hansen and Andersen (2008) found that employees with the most restrictive attitudes concerning sickness absenteeism were the most likely to show up for work while sick. In addition, it seems as if the conception of the absence culture in the workplace is significant. Bamberger and Biron (2007) suggested a conforming tendency among what they perceive as the norm in coworkers, and Johns (2011) found that the conception of an organizational culture that legitimizes sick leave is related to less presenteeism. Harvey and Nicholson (1999) found that absence legitimacy can vary in accordance with factors such as age, sex, and which sickness causes the absenteeism.

11.6 Models for Absence, Presence and Presenteeism

Some models for absence and presence have been developed in which presenteeism has been included. The first was launched in 1978 (Steers & Rhodes, 1978). Steers and Rhodes (1978) developed a model explaining an employee's presenteeism behavior. A fundamental principle of this model is that the employee's motivation is crucial for presenteeism as long as the employee has the ability to show up. This ability is influenced by, for example, sickness, accidents, family responsibility, while motivation is largely affected by a combination of the employee's affective responses to the job and the internal and external pressures to show up. Examples of the factors that could lead to influencing this pressure are as follows: the working situation, job satisfaction, working morale, job belonging, and the norms at the workplace. This model also includes attendance pressure as a term which describes the internal and external pressures to show up.

Despite the prevalence and importance of presenteeism, the field has been criticized as being largely a-theoretical (Johns, 2011). As a result, we have relatively little knowledge of mechanisms that underlie presenteeism behavior, although it should be noted this field is growing positively. Below, we outline some of the available conceptual models.

Johansson and Lundberg (2004) developed the model of illness flexibility, suggesting attendance requirements (the negative consequences that employees face due to absence) and adjustment latitude (modifying the workload of sick employees) as key determinants of sickness presenteeism and absenteeism. Aronsson and Gustafsson (2005) also suggested that there are two different types of attendance demands which influence sickness presenteeism: personal factors (such as private financial situation and individual boundarylessness) and work factors (such as control over pace of work, replaceability, sufficient resources, and time pressure and conflicting demands). Biron and Saksvik (2010) presented a taxonomy of the determinants of presenteeism which includes work-related (difficulty in being replaced), dispositional (guilt and pressure factors), and situation factors (financial insecurity). Their findings were supported by Baker-McClearn et al. (2010) who too identified workplace factors that influence an individual's decision to attend work when ill: personal motivations (e.g., loyalty to own professional image) and workplace pressures (e.g., workplace culture), and accordingly described two types of presenteeism: institutionally mediated presenteeism and personally mediated factors. Johns (2010) proposed an integrated model to explain presenteeism, according to which a "health event" triggers a choice between presenteeism and absenteeism, and that this decision also depends on the work context (ease of replacement, absence policy, job demands) and individual factors (personality and work attitudes). A later model by Gosselin et al. (2013) expanded on this research, but provided a different scope as it proposed an integrated model which identified multiple determinants of both presenteeism and absenteeism behaviors. Their model incorporated the influence of health problems, demographic characteristics, individual factors, and organizational factors on presenteeism and absenteeism. More recently, Karanika-Murray, Halley, Griffiths, and Biron (2015) viewed presenteeism as a combination of physical presence and psychological absence and demonstrated the crucial role of affective-motivational states as mediators in the presenteeism-job satisfaction relationship. Finally, Miraglia and Johns (2015) proposed a dual path model which listed as antecedents of presenteeism pressure factors (e.g., high personal or professional obligations to work) and positive motivational factors (e.g., pleasure derived from work and job satisfaction).

11.7 Further Research

Sickness absenteeism is still a popular research topic in Norway, and most of the projects deal with the causes for sickness absenteeism and how it can be dealt with and reduced. The problem with Norway and some other countries dealing with high levels of sickness absenteeism is that we have a high percentage of the population participating in the labor force, and consequently, a high level of people who have chronic sicknesses or other situations which could lead to long or repeated periods of sick leave being included in the working population. This means that there is a need for another discussion about absence reduction. As mentioned, the government in Norway has initiated graded sick leave as an initiative, and some research indicates that this slightly reduces the absence. For these initiatives that restrict the ability to take sick leave from work, we probably must facilitate the possibility of "constructive presenteeism." There is a good basis for this in Norway because of both a paragraph in The Work Environment Act as a facilitator (§ 4–2) and the contract among the work life partners of Including Work Life (IWL agreement, which was started in 2001 and prolonged in 2014–2018). However, we have little information on how the facilitation would work in action. This will probably vary depending upon what kind of job tasks the business requires and the magnitude of them. For example, there could be a greater possibility for a sick employee to continue working and become better if the individual could perform easier job tasks for a period of time. Again, one can think that this initiation might increase the workload on colleagues and therefore negatively affect the entire working environment. In understanding if graded sick leave increases in prevalence, a thorough examination of the consequences on a single workplace for an individual employee is needed to find the optimal solution for the worker's health.

References

Addae, H. M., & Johns, G. (2002). National culture and perceptions of absence legitimacy. In M. Koslowsky & M. Krausz (Eds.), *Voluntary employee withdrawal and inattendance*. New York, NY:Kluwer Academic.

Aronsson, G., & Gustafsson, K. (2005). Sickness presenteeism: Prevalence, attendance-pressure factors, and an outline of a model for research. *Journal of Occupational Environmental Medicine, 47,* 958–966. doi:10.1097/01.jom.0000177219.75677.17.

Aronsson, G., Gustafsson, K., & Dallner, M. (2000). Sick but yet at work. An empirical study of sickness presenteeism. *Journal of Epidemiology and Community Health, 54*(7), 502–509.

Baker-McClearn, D., Greasley, K., Dale, J., & Griffith, F. (2010). Absence management and presenteeism: The pressures on employees to attend work and the impact of attendance on performance. *Human Resource Management Journal, 20*(3), 311–328.

Bamberger, P., & Biron, M. (2007). Group norms and excessive absenteeism: The role of peer referent others. *Organizational Behavior and Human Decision Processes, 103*(2), 179–196. doi:10.1016/j.obhdp.2007.03.003.

Biron, C., Brun, J.-P., Ivers, H., & Cooper, C. L. (2006). At work but ill: Psychosocial work environment and well-being determinants of presenteeism propensity. *Journal of Public Mental Health, 5*(4), 26–37. doi:10.1093/eurpub/ckp076.

Biron, C. & Saksvik, P. Ø. (2010). Sickness presenteeism and attendance pressure. Factors: Implications for practice. In C. L., Cooper, J. C. Quick, & M. Schabracq (Eds.). *Work and Health Psychology. The Handbook* (3rd ed.). London: John Wiley & Sons, Ltd.

Bjørgum, H., Kristoffersen, H. B., & Kalajdzic, P. (2012, May 22). Du blir friskere av å bruke gradert sykemelding. *NRK*. Retrieved from http://www.nrk.no.

Böckerman, P., & Laukkanen, E. (2010a). Predictors of sickness absence and presenteeism: Does the pattern differ by a respondent's health? *Journal of Occupational and Environmental Medicine, 52*(3), 332–335. doi:10.1097/JOM.0b013e3181d2422f.

Böckerman, P., & Laukkanen, E. (2010b). What makes you work while you are sick? Evidence from a survey of workers. *European Journal of Public Health, 20*(1), 43–46. doi:10.1093/eurpub/ckp076.

Claes, R. (2011). Employee correlates of sickness presence: A study across four European countries. *Work & Stress, 25,* 224–242. doi:10.1080/02678373.2011.605602.

Collins, A., & Cartwright, S. (2012). Why come into work ill? Individual and organizational factors underlying presenteeism. *Employee Relations, 34*(4), 429–442. doi:10.1108/01425451211236850.

Cooper, C., & Dewe, P. (2008). Well-being—absenteeism, presenteeism, costs and challenges. *Occupational Medicine, 58,* 522–524.

Dahl-Jörgensen, C., Opdahl Mo, T. & Saksvik, P. Ø. (2002). Utvidet egenmelding som strategi for å bedre subjektiv helse og redusere sykefravær [Enhanced self-certification as a strategy to better subjective health and reduce sick leave]. *Norsk Samfunnsvitenskaplig Tidsskrift, 43,* 3–29.

Elstad, J. I., & Vabø, M. (2008). Job stress, sickness absence and sickness presenteeism in Nordic elderly care. *Scandinavian Journal of Public Health, 36*(5), 467–474. doi:10.1177/1403494808089557.

Gosselin, E., Lemyre, L., & Corneil, W. (2013). Presenteeism and absenteeism: Differentiated understanding of related phenomena. *Journal of Occupational Health Psychology, 18*(1), 75–86.

Grinyer, A., & Singleton, V. (2000). Sickness absence as risk-taking behaviour: A study of organisational and cultural factors in the public sector. *Health, Risk & Society, 2*(1), 7.

Grødal, K. (2014). *Syk, men likevel på jobb* [Sick, but still at work] *(Master thesis)*. Dep. of Psychology, NTNU, Trondheim.

Hansen, C. D., & Andersen, J. J. (2008). Going ill to work—What personal circumstances, attitudes and work-related factors are associated with sickness presenteeism? *Social Science and Medicine, 67,* 956–964.

Hansen, C. D., & Andersen, J. H. (2009). Sick at work–a risk factor for long-term sickness absence at alater date? *Journal of Epidemiology and Community Health, 63*(5), 397–402.doi:10.1136/jech.2008.078238

Harvey, J., & Nicholson, N. (1999). Minor illness as a legitimate reason for absence, *20,* 979–993.

Heponiemi, T., Elovainio, M., Pentti, J., Virtanen, M., Westerlund, H., Virtanen, P. … Vahtera, J. (2010). Association of contractual and subjective job insecurity with sickness presenteeism among public sector employees. *Journal of Occupational and Environmental Medicine, 52*(8), 830–835. doi:10.1097/JOM.0b013e3181ec7e23.

Johansson, G., & Lundberg, I. (2004). Adjustment latitude and attendance requirements as determinants of sickness absence or attendance. Empirical tests of the illness flexibility model. *Social Science and Medicine, 58*(10), 1857–1868.

Johns, G. (2010). Presenteeism in the workplace: A review and research agenda. *Journal of Organizational Behavior, 31,* 519–542. doi:10.1002/job.630.

Johns, G. (2011). Attendance dynamics at work: The antecedents and correlates of presenteeism, absenteeism, and productivity loss. *Journal of Occupational Health Psychology, 16,* 483–500. doi:10.1037/a0025153.

Karanika-Murray, M., & Biron, C. (2016). *Presenteeism can be functional for health and performance: A critical review of the literature with recommendations for research.* Manuscript submitted for publication.

Karanika-Murray, M., Pontes, H. M., Griffiths, M. D., & Biron, C. (2015). Sickness presenteeism determines job satisfaction via affective-motivational states. *Social Science and Medicine, 139,* 100–106.

Lie, S. A. (2009). Evaluering av tiltak for å redusere sykefravær [Evaluation of means for reducing sickness absence]. *Norsk Epidemiologi, 19*(2), 153–160.

Lindberg, P., Josephson, M., Alfredsson, L., & Vingård, E. (2008). Promoting excellent work ability and preventing poor work ability: The same determinants? Results from the Swedish HAKuL study. *Occupational Environmental Medicine, 63,* 113–120.

Löve, J., Grimby-Ekman, A., Eklöf, M., Hagberg, M., & Dellve, L. (2010). "Pushing oneself too hard": Performance-based self-esteem as a predictor of sickness presenteeism among young adult women and men—A cohort study. *Journal of Occupational and Environmental Medicine, 52*(6), 603–609. doi:10.1097/JOM.0b013e3181dce181.

Lund, R., Rognstad, K., Rundmo, T., & Saksvik, P. Ø. (1991). *Kostnader ved arbeidsulykker og yrkesrelatert helsesvikt* [*Cost related to work accidents and work related health problems*] (Research report STF83 F91001). Trondheim: SINTEF, Inst. for anvendt økonomi.

Markussen, S., Mykletun, A., & Roed, K. (2012). The case for presenteeism—Evidence from Norway's sickness insurance program. *Journal of Public Economics, 96*(11–12), 959–972. doi:10.1016/j.jpubeco.2012.08.008.

McKevitt, C., Morgan, M., Dundas, R., & Holland, W. W. (1997). Sickness absence and "working through" illness: A comparison of two professional groups. *Journal of Public Health Medicine, 19*(3), 295–300.

Milch, V. (2011). *Sykenærvær—et forsømt tema?: En kvantitativ undersøkelse av sykenærvær i en IA-virksomhet [Presenteeism—a neglected topic?: A quantitative study of presenteeism in an IW Enterprise] (Master Thesis).* Trondheim: Norges Teknisk- Naturvitenskapelige Universitet.
Miraglia, M., & Johns, G. (2015). Going to work ill: A meta-analysis of the correlates of presenteeism and a dual-path model. *Journal of Occupational Health Psychology, 20*(1), 1–14. doi:10.1037/a0037754.
Rostad, I. S., Milch, V. & Saksvik, P. Ø. (2015). Psychosocial workplace factors associated with sickness presenteeism, sickness absenteeism, and long-term health in a Norwegian industrial company. *Scandinavian Psychologist, 2*, e11. http://dx.doi.org/10.15714/scandpsychol.2.e11.
Saksvik, P. Ø. (1996). Attendance pressure during organizational change. *International Journal of Stress Management, 3,* 47–59.
Simpson, R. (1998). Presenteeism, power and organizational change: Long hours as a career barrier and the impact on the working lives of women managers. *British Journal of Management, 9,* 37–50. doi:10.1111/1467-8551.9.s1.5.
Steers, R. M., & Rhodes, S. R. (1978). Major influences on employee attendance: A process model. *Journal of Applied Psychology, 63*(4), 391–407.
Thun, S., Saksvik, P. Ø., Ose, S. O., Mehmetoglu, M., & Christensen, M. (2013). The impact of supervisors' attitudes on organizational adjustment norms and attendance pressure norms. *Scandinavian Journal of Organizational Psychology, 5,* 15–31.
Wynne-Jones, G., Buck, R., Varnava, A., Phillips, C., & Main, C. J. (2009). Impacts on work absence and performance: What really matters? *Occupational Medicine, 59,* 556–562.

Chapter 12
Healthy Change in Intervention Research and Reorganization

Per Øystein Saksvik and Maria Karanika-Murray

In this chapter, we will look at reorganization and interventions that aim to improve the conditions for employees to enable them to work under more health-promoting conditions. In this respect, we will examine the common core of two different research traditions: intervention research and research on organizational change. However, these traditions are often highly parallel, so much so that we sometimes wonder why they are not viewed more in context. Either way, to begin with we choose to keep the traditional separation and will begin by considering interventional research.

12.1 Present Directions Within Interventional Research

With a view to increasing working health and well-being, many organizations carry out interventions at regular intervals. This does not mean that it is easy to succeed with such interventions, at least not when assessed in terms of what the impact of the intervention was thought to be. An intervention is a systematic measure directed toward a goal. There are different kinds of interventions, and they can be directed toward different levels of an organization. This section of the chapter will primarily be concerned with two aspects of interventions: (1) compensating interventions, which are proactive and promote health measures that highlight the positive aspects of the working situation and, through this, counteract the effect of negative situations and events, and (2) the implementation of the intervention (the process),

P.Ø. Saksvik (✉)
Department of Psychology, Norwegian University of Science and Technology (NTNU), Trondheim, Norway
e-mail: per.saksvik@ntnu.no

M. Karanika-Murray
Department of Psychology, Nottingham Trent University, Nottingham, UK
e-mail: maria.karanika-murray@ntu.ac.uk

M. Christensen et al. (eds.), *The Positive Side of Occupational Health Psychology*,
https://doi.org/10.1007/978-3-319-66781-2_12

which refers to how different interventions are being introduced and the implementation practice used. Considerable research is being conducted in both areas, the aim of which is to contribute to the enhancement of valuable interventions introduced into organizations.

It has long been an established truth that interventions that directly target the reason for a problem are the best, as well those that obtain the greatest effects by initiating interventions at the organizational level (LaMontagne, Keegel, Louie, Ostry, & Landbergis, 2007; Nielsen & Randall, 2012). Therefore, substantial resources are being used to uncover the causes of such problems through, for example, the use of questionnaires. The question is does it pay off to uncover the causes, especially within a multifaceted organization where the causes may be highly complex. Furthermore, the aim here is to take a closer look at the different categories of interventions, and especially the new category of countervailing interventions.

There are different kinds of interventions, and they can be directed at different levels. The main clusters of interventions we operate with the aim of changing or preventing are termed primary, secondary, and tertiary interventions (Reynolds, 1997; Richardson & Rothstein, 2008). The usual level of operation is with the single employee, a department, or a group of employees; alternatively, they may aim to introduce change at the organizational level (Saksvik & Nytrø, 2005). It is also possible to introduce interventions on a community or a national level, or a combination of these.

12.2 Primary Interventions

The first level, primary interventions, includes activities that aim to remove, change, or avoid the casual factors behind the problems. A visible sign of an underlying problem can be a marked increase in sickness absenteeism or a steep rise in an organization's staff turnover. When one or more of these causes are revealed, an intervention at an individual level may involve, for example, better training for individuals with special working conditions or needs (Saksvik & Nytrø, 2005). This could, for example, include organizing the work so that the employee, who otherwise would be on a sick leave, can do their job as close to normal as possible (Biron & Saksvik, 2010; see also Chap. 6). On a department/group level, measures can include facilitating positive working habits and contributing through socialization measures and training, such as introductory courses for the employees (Saksvik & Nytrø, 2005). Examples of measures at an organizational level can be to reduce job demands, clarify the leadership role, improve poor turnover arrangements, increase possibilities for social support, facilitate career development, or increase the employees' sense of control and codetermination (Dunham, 2001; Reynolds, 1997; Richardson & Rothstein, 2008; Saksvik and Nytrø, 2005).

12.3 Secondary Interventions

Secondary interventions include activities that aim to reduce the severity of symptoms before they develop into more lasting health issues (Reynolds, 1997; Richardson & Rothstein, 2008). On an individual level, one can carry out intervention programs that involve stress management or creating training facilities (Saksvik & Nytrø, 2005). The aim might be to create better conditions for those in the "danger zone" to increase their stress management or enhance their health in general (Reynolds, 1997; Saksvik & Nytrø, 2005). At a group/department level, measures may be introduced aimed at risk groups, for example, those with demanding and monotonous job situations (Saksvik & Nytrø, 2005). On an organizational level, these might include regular surveys of potential demands in the organization (Saksvik & Nytrø, 2005). In today's working life, secondary interventions are those most commonly used, and the intention is to prevent or reduce factors that can contribute to poorer health or lower productivity among employees (Saksvik & Nytrø, 2005). Examples can be cognitive-behavioral skill training, relaxation, meditation, physical exercise, time management, and goal setting (Richardson & Rothstein, 2008).

12.4 Tertiary Interventions

Tertiary intervention measures address sickness (disorders) and problems, with the aim of hindering or restricting the problem and its effect on the practice of individuals (Reynolds, 1997; Richardson & Rothstein, 2008). Targeting individuals can mean providing help to those who show clear signs that stress factors are beginning to have consequences. Such interventions can also include rehabilitation after sickness absenteeism (Dunham, 2001). On a group level, measures may include services such as counseling and psychotherapy, or support measures with qualified health professionals available in the workplace (Reynolds, 1997; Richardson & Rothstein, 2008). Other measures can target specific groups, for example, those with chronic diseases (Saksvik & Nytrø, 2005). Tertiary prevention on an organizational level can involve the business creating good routines for contact with, and follow-up of those on sick leave (Saksvik & Nytrø, 2005).

12.5 Community-/National-Level Interventions

There exists a fourth level of implementation, one that takes place at the community or national level. Such measures may include the introduction of regulations on internal control, as occurred in Norway in 1992 and, in 1997, these were converted into regulations on systematic health, environment, and safety (Saksvik & Nytrø, 2011).

These regulations require Norwegian businesses, independent of the number of employees and industry, to implement the following measures. Businesses must undertake proper systematic actions to ensure that the enterprise operates in accordance with the requirements specified in law and regulations in the domains of health, environment, and safety. In the original regulation from 1992, great emphasis was placed on the documentation, meaning that the business could present their system (rules, strategies, etc.) in a format that could be inspected by the supervisory authorities without necessarily seeing the system at work in practice. The documented system was designed as a handbook that represented the different aspects of the system. This demand was toned down in the revised regulation in favor of "activities," that is, OHS measures that were actually accomplished. Implementing this system in a business can be viewed as an intervention in itself, but only when the system leads to concrete measures that target individuals, groups, or the organization as a whole, where the intervention can be thought of as active. This level of interventions are also very good supplementary measures to more focused interventions. So, they may be useful to kick-start change and organizations' willingness to act, at least at the start. This was the case in the Netherlands who introduced something very similar to Britain's management standards for stress at least a decade before Britain did. In this case, regulation helped to kick-start organization actions in this field.

12.6 Countervailing Interventions

In recent years, a fourth type of intervention has arrived with full force. The term "countervailing" is being used when explaining these kinds of interventions (Kelloway, Hurrel, & Day, 2008), which include proactive and health-promoting measures that highlight the positive aspects of the working situation and, through this, counter the effect of negative situations and events (Milch, Vaag, Giæver, & Saksvik, 2013). A significant aspect of this kind of intervention is that they are not related to the causes of eventual problems. The aim is to strengthen individuals and organizations through health-promoting activities. One example of this kind of activity is the project "The sound of well-being" that was carried out in a number of organizations including one department of a health institution (Vaag, Saksvik, Theorell, Skillingstad, & Bjerkeset, 2012). The main content of this intervention was to establish a choir in the units or departments. The choir practiced with professional instructors outside of working hours and finally gathered on a stage where the choirs competed against each other. The results from this intervention showed that more women than men participated and that in contrast to non-participants, the participants reported more engagement, a greater sense of belonging to the organization, and a better psychosocial working environment after the intervention (Vaag et al., 2012). In this example, the intervention was directed toward improving well-being in the organization, making it more attractive and meaningful, increasing the organization's reputation, attracting new employees, and

Level/ type	Individual	Group	Organization	Society
Primary	Adjusting work for positive presenteeism			
Secondary		Education in Stress management		
Tertiary			Follow-up of long term sickness absentees	Systematic OHS-work
Compensating (Countervailing)			«Sound of well-beeing»	

Fig. 12.1 Classification of interventions

reducing staff turnover. Seen as such, this intervention intended to influence several levels. Although, the initial location was at an organizational level, both individuals and working groups gained from participation.

The intervention presented here can therefore be organized in the Fig. 12.1: [For more examples on the first three levels, see LaMontagne et al. (2007)].

Considerable research has shown that interventions at the organizational level are most effective (Nielsen & Randall, 2012; LaMontagne et al., 2007). This is probably because the measures must be implemented in complex organizational settings. For the solutions are to have an impact and become permanent, they must be embedded at the level where decisions about business strategies are being made. Still, it is possible to achieve effects on a lower level as well. If some employees feel as if they are under a lot of pressure, one might review the working procedures of those who are affected and make changes, for example, by allocating the workload better, by increasing the number of staff, or by offering techniques of stress management to those with higher workloads. However, it is difficult to see such measures as completely separate from the business. Most of the measures on an individual or relational level have implications for the organizational level. Increasing the number of staff and resolving unbalanced workloads often have large implications for the whole working organization and will affect the experience of justice in the whole organization. If someone receives an offer of education in stress management, this may be experienced as stigmatizing both for those who receive the offer and those who do not.

12.7 The Interventional Process

In a presentation in Nottingham in 2012, one of the authors of this book suggested that one might use the analogy of a troll to explain the difficulty of functioning interventions (Saksvik, 2012). One of the unruly and dangerous trolls mentioned in that presentation was related to how an intervention is implemented, and this was termed the interventional *process* (Fig. 12.2).

Over a long period, our research has focused on what it takes for an intervention to be successful, and we have claimed that the process is more important than its content (troll 1). In an earlier summary of the literature on successful restructuring processes in an organization, we concluded that there was a lack of insight into reactions to change—in the changing processes (see Nytrø, Saksvik, Mikkelsen, Bohe, & Quinlan, 2000; Saksvik, Nytrø, Dahl-Jørgensen, & Mikkelsen, 2002). One of the earlier observed issues, both during and after a planned change process, was that there is a diffuse resistance against changes that can hinder goal achievement. In some cases, this resistance may be understandable or it may be perceived as hidden or open "sabotage" depending on which group one identifies with within the business.

The literature dealing with change in an organization prescribes ideal ways to change the business but seldom describes the actual change process in a business that has not managed to meet the already accepted goals. Documented success stories exist about businesses that have succeeded and what one can learn from that (e.g., Peters & Waterman, 1982), but it is equally important to extract lessons from organizational developmental processes that have not succeeded. An important factor is to seriously consider that local development strategies based on unilateral approaches (top-down changes) are more likely to fail than those that use participation strategies, which also safeguard the organization's resources. Process thinking that is linked to interventional research is, in other words, at the center of a model for Norwegian working life where participation and cooperation are the keys.

A current issue associated with an interventional process is the question of how this process can be evaluated. This has been so unclear that in recent articles, the evaluation process has been called the black box (Nielsen & Randall, 2012). However, we have conducted several studies of organizational change in general that can contribute some clarifications (these will be addressed in the second part of this chapter) (Armenakis & Bedeian, 1999; Burke, 2014; Tvedt & Saksvik, 2012; Lien & Saksvik, 2016). When we conducted an assessment of an occupational

Fig. 12.2 *Source* Theodor Kittelsen: «Trollet som satt og grunnet på hvor gammelt det er» ("The troll who sat and pondered how old it is") (1911)

health project in a municipality (that was completed in the mid-1990s), we used a relatively open qualitative approach (Dahl-Jørgensen & Saksvik, 2005) but, more recently, questionnaires have also been used (Randall, Nielsen, & Tvedt, 2009; Tvedt, Saksvik, & Nytrø, 2009; Saksvik, Olaniyan, Lysklett, Lien, & Bjerke, 2015; Abildgaard, Saksvik, & Nielsen, 2016). Researchers evaluating the implementation part of organizational interventions have previously used a simple question about process (Murta, Sanderson, & Oldenburg, 2007). The most common aim of such evaluations has been the participant's exposure level to the intervention. The findings from these studies, which have compared the participant's exposure to interventions, have been inconsistent, with some reporting positive health outcomes with increased exposure (Elo, Ervasti, Kuosma, & Mattila, 2008; Murta et al., 2007; Randall, Griffiths, & Cox, 2005) and others finding no evidence for the same effects (Arnetz, 1996; Wilson et al., 2010). Studies that compared participants with non-participants in the same organization are rare, but one study showed that sex, personality, and some of the working conditions differed between the two groups (Vaag et al., 2012). Other studies have examined the participant's perception of implementation in relation to their own health outcomes (Nielsen, Randall, & Albertsen, 2007; Randall et al., 2009). For example, the employee's experience of the general manager's attitudes and behavior were found to be more important for their future actual health than their exposure to the actual intervention, which in this case was team exercise (Randall et al., 2009). The employee's perception of information and communication concerning the intervention has also been related to the interventional effect (Jimmieson, Terry, & Callan, 2004). Employees with a higher level of participation in the change process have shown less resistance toward change and a greater achievement of goals and organizational belonging (Lines, 2004). Equally, high levels of participation in the change process have been associated with lower levels of stress symptoms and higher job satisfaction in post-measurements (Nielsen et al., 2007), less self-reported job demands, an increase in social support, and a decrease in the experience of stress (Eklöf, Ingelgård, & Hagberg, 2004).

Our final contribution to understanding the implementation process derives from research that we conducted in a project that was carried out in an operating and accounting department that introduced something they called a co-worker program. The aim of the program was to enhance the working environment in the department. In this study, we found two significant process factors: communication and leadership. These factors could explain most of the satisfaction with the co-worker program (Saksvik et al., 2015).

Another troll that stands in the way of interventional success is whether the intervention has a natural foundation in the business (troll 2). We researchers have a preconception of us being in the service of good and as the bearers of brilliant proposals, and there is no reason why the intervention should not be implemented. We thus have a "bezzerwizzer" and external approach, which is not necessarily accepted by either the staff or the management of the business. One way forward is to work thoroughly on the preparation, which requires taking a great deal of time and energy. Sometimes, in our own research we have tried an alternative approach that consists of "connecting with" the already initiated interventions in the organization. We term these as natural interventions (Kico & Saksvik, 2015). Of course, in choosing this strategy, researchers lose a great deal of the control that they would otherwise have over the intervention in terms of content, design, and implementation. What one gains is the inner motivation to succeed in the business, at least if the parties agree on testing the measure. As an example of such a natural intervention, we chose "experiments with expanded self-certification of sickness absenteeism" (Saksvik & Nytrø, 2001, p. 83)—an intervention that was implemented in an organization (municipality/local council) shortly before the new millennium. The intervention simply consisted of the opportunity for employees to increase the number of self-certification days of sickness absenteeism to five. The limit had previously been three days. The number of times a year that self-certification could be used—five times—was unchanged. Fortunately, we additionally got a good research design, since the municipality did not implement the program in the whole municipality, but only in a limited part. We were then able to compare the trial with the other part of the municipality for over a year. We completed different evaluations (questionnaire and interview), as well as followed the sickness absenteeism statistics (Dahl-Jørgensen & Saksvik, 2002; Saksvik & Nytrø, 2001). A number of interesting observations emerged: Relatively few employees made use of the program; mild and subjective health issues formed the basis of more frequent use of self-certification; and there were clear barriers in the absenteeism culture of the organization that determined how easy or difficult it was to adopt such a program (such as consideration for the patients/users or the fear of management suspecting them of being a shirker). In other words, for such an intervention to succeed, it was important to have good flexibility in the organization.

All in all, we doubted whether extended self-certification was a good enough "medicine" for preventing increases in sickness absenteeism, but in retrospect, we have seen that extended self-certification—up to eight times per year and a total of 24 days—is recommended in the agreement for an inclusive workplace in Norway (The IA agreement). This intervention has become political and is part of a social intervention that may not really have a foundation based on research in relation to whether the measure has an impact. Nevertheless, we agree that expanded self-certification is positive in that it acts as a sign of trust in the employee's evaluation of their own health condition as a basis for sickness absenteeism. There are, however, too many aspects of such an intervention that have not yet been thoroughly examined. This is also a good example of the Norwegian model where control regimes are torn down on the basis of great trust between the individual employee and the employer. Simultaneously, however, it becomes possible to debate criticism of this model as being naïve and easily abused. There will always be one or two individuals who see the self-certification program as a rightful holiday and will make use of the maximum amount of days.

12.8 Healthy Change

When a company or a public business goes through reorganization, occupational health becomes less of a priority. The aim is first and foremost a better and increased effective organization that produces more or better products and services. For over a decade, some of the authors of this book have been working with a healthy perspective on restructuring or reorganization. The aim of this part of the chapter is to provide an overview of our results. Over a three-year period, almost a third of Norwegian employees reported change in their working situation and fewer than half reported this to have an effect on their working day (Aagestad et al., 2011). From the media, we get an impression that restructuring is a significant part of our everyday job, as seen in the hospital reforms in 2001, the restructuring in SAS with the transition to national companies in 2004, the merger of Statoil and Hydro in 2007 and Mamut and Visma in 2011, and large-scale reorganization in the postal services both in 2003 and in 2010 (Orucevic, 2013). Restructuring is essential in today's working life since it provides opportunities for growth, development, and increased resources for companies, and a successful change can be the key to realizing a company's potential in the market (Lewis, 2011). Nevertheless, about 80% of all restructuring fails (Nguyen & Kleiner, 2003; Orucevic, 2013).

Ten years ago, our belief was that restructuring is more effective and efficient when the aspects of occupational health are safeguarded. However, in the preparation of our big restructuring project (see, e.g., Saksvik et al., 2007), we quickly gained an impression of restructuring as being difficult and, as noted above, as easily open to failure. It is simply the case that the majority of initiated organizational changes in businesses and companies do not achieve the goals they desired (Nytrø et al., 2000). This may be because they are too ambitious, the results are not immediately present, or that it is difficult to evaluate the effects. Our hypothesis, however, was that there is a lack of emphasis on the implementation of change and that this explains the modest results. If one accounts for the implementation process, the occupational health aspects are automatically included because of the human factor in the calculation. When the employees are properly included in the change, this will become easier (Lien & Saksvik, 2016).

Our thinking is consistent with the new Working Environmental Act in Norway, which was implemented in 2006. In the Act, there is a separate section on requirements for a good restructuring process (article 4.2) which states explicit requirements for information, participation, and skill development. These requirements should be viewed as ideal, and they involve how the business structures change. Thus, for many this section of the act became more an organizational principle rather than a reality that has less direct interaction with occupational health-related realities. If occupational health was the reason for restructuring, it would be natural to look at the relationship between (re)organization and job uncertainty.

Each change, small or large, has an impact on the employee's everyday work. Downsizing naturally entails a greater job uncertainty than does a redefinition of a job position (e.g., new tasks, new structures in the position, new working team), but still this is not necessary. Each individual reacts differently to these situations. As summarized by Østerud (2014), research on job uncertainty reveals numerous negative psychological and organizational consequences. People, who experience a high level of job uncertainty, more often have plans to terminate their employment (O'Driscoll & Beehr, 1994; Rafferty & Griffin, 2006), or it leads to other psychological reactions such as emotional drain (Paulsen et al., 2005) and decreased job satisfaction (Nelson, Cooper, & Jackson, 1995; Paulsen et al., 2005; Rafferty & Griffin, 2006). Furthermore, it has been found that the experience of job uncertainty can contribute to increased stress (Pollard, 2001) or to a generally increased risk of poorer health (Kivimäki et al., 2001; Maurier & Northcott, 2000), increased blood pressure (Pollard, 2001), anxiety (DiFonzo & Bordia, 2002), and reduced job performance (Orpen, 1994).

To avoid employees ending up in an unsafe situation when restructuring, there are many activities that can be undertaken as preventative measures. Around the new millennium, our research group at NTNU began our project on "Healthy Organizational Change" (HOC) as the basis for our research. Our perspective on healthiness was inseparable from employee health, and we were especially interested in the way in which a restructuring was carried out, that is, the implementation process. Our first review of the literature and our empirical studies taught us about the different phases of change, and which criteria must be met for the change to be successful (Nytrø et al., 2000; Saksvik et al., 2002). Essentially, our first impression was that because change is so complicated to implement, it should be thought through carefully before beginning the process. When we then learned that some companies and businesses reconstruct almost as a strategic move, without sufficient reasons for doing so, it was obvious to us that someone is taking an unnecessary risk. Yet, even when change forces its way, it is also difficult to determine when change is needed. It is easy to interpret the signals from the market or internal development too late, so that so-called sudden changes become necessary (Burke, 2014). Burke uses the term "revolution" when explaining what happens in these cases, and with such a strong notion, it is evident that change can be demanding. If we were going to use our own suggested model of occupational psychology (see Chap. 3), in these situations we are not talking about simple tweaks and adjustments in line with an understanding of balance and harmony. The forces are against each other in a learning and developmental process that can be so demanding that expertise is not always sufficient for achieving a good change.

In our HOC project, we quickly learned what it would take to achieve healthy change. Our first empirical study was based on 91 companies and businesses undertaking reconstruction (Saksvik et al., 2007). We interviewed one representative for the leaders and one for the employees in cooperation with a representative from the labor inspection, 180 interviews in total. The method of analysis was based on Grounded Theory, which is a systematic way to collect and process qualitative (interview) data and develop theory. Our study showed that the meaning of "knowing oneself" prior to change, that is, insight into the businesses norms and diversity of reactions, and the meaning of a present leadership, clarifications of roles and handling of conflicts constructively can increase the likelihood of the changes success, is important (Saksvik et al., 2007). Some of these factors may appear easy to address. It may seem that identifying "resources" from the JD-R model (see Chap. 3) is important, that is, making sure that during and after a restructure, one has the necessary resources needed to maintain the normal operation.

Similarly, it is easy to think of "social support" as a central concept and for that reason leadership is presented as an important factor. Yet, it is more fundamental factors in an organization that must also be considered. This can be illustrated with an adjoining project to the HOC project in which we studied leaders who had succeeded in introducing change (as assessed by the outside world of the business). In this project, which was entitled "Good leaders in the public sector," we found a further emphasis on how good leaders work during the implementation of change. A key finding here was that many executives apparently exhibit great tolerance for the uncertainty that follows change (Øyum, Andersen, & Saksvik, 2006). They usually appeared as "shock-absorbers" for the employees and were still able to prioritize daily operations while the change was ongoing. The employees' felt confident that the leaders knew how to maneuver in a similarly chaotic situation and they utilized shared values as a tool for management. For several of these leaders, it was also clear that they needed the employees' engagement and assistance in the change process (Øyum et al., 2006).

The classical example in this context was the leader who always had the office door open and as a result had to set up the name plate on the wall so that people could find the office. The important role of leaders in the implementation of change is reflected in the work of Saksvik and Tvedt (2009) who portrayed this as a CEO who wishes to ensure that the employees in her working group have access to adequate information regarding the organization's development and future, and to reassure them that they share common values, but at the same time inspire them to challenge and change these values. A leader should support a culture of experimentation, be open, and learn from mistakes and misunderstandings. Thereby the employees can move in new directions, challenge the basic values embedded in the organization, and develop an insight into change as unpredictable and sometimes irrational. One of the biggest reasons for why changing organizations fail in achieving their goals is the lack of openness to change, and leaders are essential for that.

12.9 Change Models and Processes

Many realize that organizational change is hard and throw themselves into a market of expert consultants whose role is to assist with such change. The latest contribution on this front offers lectures from the celebrity lawyer Geir Lippestad (Publicom Forlag, 2014). From an advert, it appears that value-based communication is at the core of these lectures. The lectures may well be inspirational, but the core of how change takes place has been known for a long time. One of the biggest names in organizational change was already active in the 1930s, launching his "recipe" for change in 1951 (Lewin, 1951) (Fig. 12.3).

The classical model of change, which is one of the best known in the psychological literature in general, is Kurt Lewin's Three Step Model from 1951. Based on this model, today's practice first has to "unfreeze," opening up the potential for finding alternative solutions. The second step is termed "change," where we see a systematic change process before the new structure becomes a permanent part of the

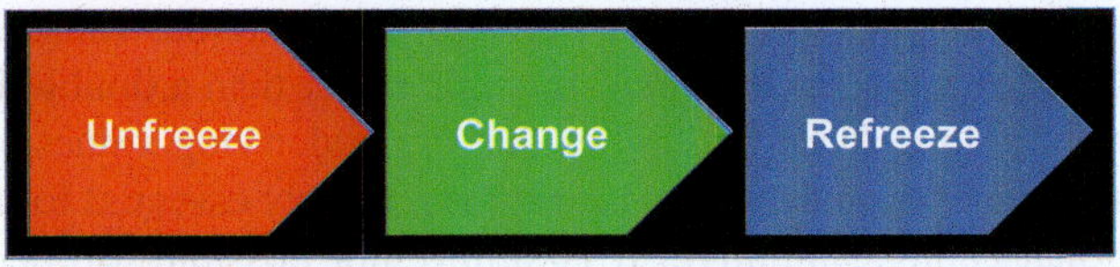

Fig. 12.3 Kurt Lewin's change model

organization's way of reacting and acting. An organization will always protect itself from individual resistance to change and group conforming behavior. Dissolving established behavior patterns can occur in three operations. First, it can be achieved by increasing the forces that move the behavior away from existing situations or status quo. Second, it can limit the forces that restrain movement away from the status quo, and third, one can find a combination of these two methods (Kritsonis, 2005). Kritsonis (2005) states that the kinds of activities that are usually included in the change process are motivation of the co-workers, explaining the need for change, building trust, and making use of group brainstorming as a tool for problem solving. The third step, "refreeze," is about making the change permanent. For this, it is necessary to continually formalize and reinforce the new behavior in the organization through formal and informal mechanisms, that is, through regulations, policy documents, and procedures (i.e., refreeze or really just freeze) (Lewin, 1951). Lewin's change recipe has great appeal, perhaps because it is a simple and comprehensive model. The model is usually placed in the category of "planned change" (Burke, 2014). Lewin postulated a form of balance theory in which the unbalance that was created in Step 1 (unfreeze) moved the organization into a new stadium in Step 2 (change). Here, it is crucial to convince the employees that change will serve them, and to encourage them to seek new information and to cooperate and associate with leaders who have confidence in the organization and support change. Whether it actually is a balance theory, is a little problematic (see also Chap. 3). It is possible that there exists a balance before and after change, but the aim of this thinking is that there will be a change phase in between. It is in this change phase that something qualitatively new happens. As mentioned above, it is important to convince the employees about the advantages of change and for leaders who have confidence in the organization to be in the front seat when change is implemented. This has to some extent been confirmed through our HOC project and the adjoining "Good leader" project. But do we have insight into what happens in the qualitative change process?

It is possible to gain an insight into what goes on in this kind of change process by turning to another classic within organizational learning and change that of Argyris and Schön (1996). Without attempting to reproduce all their contributions to understanding organizational learning, it is important to emphasize their concept of single- and double-loop learning. The difference between these two learning forms is essential when understanding healthy change. Sometimes (perhaps usually), we solve problems by adjusting our operations in accordance with current norms and beliefs. In this case, there will be no change to the values that underpin our actions. In other words, we conduct a single rather than a double-loop learning. In contrast, double-loop learning implies that the underlying values and norms are modified. This type of organizational hearing is probably essential for the success of an organizational change initiative. It is simply not possible to operate the organization under the same assumptions as before. This is perhaps easier to understand when the whole foundation of the corporate business changes as a result of the product or service no longer being relevant to the market as a consequence of newer and/or better options. This happened when the CD was invented, but even more so when music began to be conveyed through streaming services. It becomes difficult to continue with the traditional record stores, although a few have survived for purely nostalgic reasons, or perhaps because of the sound quality of LPs that are still sold over the counter.

Less dramatic change can lead to the need to think differently. A common cause of change is caused by an attempt to increase production. In our study of an organizational reorganization program, the change was initially simple: A six-hour working day was introduced (Kico & Saksvik, 2015). A two-hour reduction of the working day would be a simple adjustment, both for the business and the employees who would retain their salary. But other conditions were also present. The change was justified by the acquisition of new technology to streamline production, which was fish processing. New machines and assembly lines led to an increased throughput of fish. By making use of the new (and expensive) equipment, they also had to introduce a two-shift arrangement. This changed the whole foundation of the company; it was a completely new way of working, which among other things resulted in greater speed and fewer breaks for the employees. The basis of the employee's health was clearly affected, and the individual impact could be positive or negative depending on things such as the kindergarten opening hours, an individual need for a positive social environment at work, or the possibility of recovery during leisure time.

In order for the organization's existence/performance is justified/optimized, it is important to experiment with new ideas. This is the same principle that was shown in conjunction with chaos theory, as noted in Chap. 3. Through their action learning reoriented thinking on learning loops, Argyris and Schön (1996) have suggested that one does not need to use a fail-and-learn strategy but, rather, can use critical reflection as a method for changing behavioral theory (Finger & Asún, 2000, pp. 45–6). Through double-loop learning, it is possible to adjust one's way of understanding practice. Finger and Asún (2000) highlight an important element that has to be present in this process of reflection: A facilitator should be present to contribute as a teacher, coach, or mentor. This facilitator can of course be close to a leader, as we saw of some leaders in the "Good leaders" project. However, it is not given that all leaders should take on this role. A coach or mentor might possess knowledge that may be useful to the learner. A middle manager is often in the same situation as a regular employee, and therefore when it comes to the meaning of change, he or she knows no more than the employee. It is essential to be open to change, to be eager to learn, and to be critical and ask questions. These elements can be perceived as properties one is born with, but from our perspective, there is also a question of attitudes that can be developed in the individual. As Bateson suggested, what is required here is a kind of meta-learning or Deutero learning (Bateson, 1979; Visser, 2002). Some theorists have termed this the triple-loop learning (Tosey, Visser, & Saunders, 2012. This term is basically about learning how to learn, that is, the ability to learn new information. This can be seen as a method that the individual can use to reorient him/herself, in addition to having a basic positive attitude toward seeing opportunities in terms of their own job and business. Job crafting can be viewed as a tool for thinking in new ways in terms of their own jobs. Job crafting is about the responsibility of employees to match the job with their own values, strengths, and passions (Wrzesniewski & Dutton, 2001; see also Chap. 7). A more specific definition is that the individual changes the job demands and job resources (Tims & Bakker, 2010). If both the employee and the manger see the opportunities in these situations, this is an example of a positive change process.

12.10 The Future

Organizational change and interventions for improving the working conditions are difficult, and if the result is to be a good working health, several conditions must be met. As summarized by Lien and Saksvik (2016), change can take many forms and will exist in both time and space, where space represents what exactly is to be changed and time refers to the process by which the change will be managed (Burke, 2014; Saksvik et al., 2007). Burke (2014) also mentions that the content of the change will, for the most part, be the management's decision. The process, however, is highly dependent on employees' involvement, commitment, understanding, and support for the change (Burke, 2014). Indifference to the process can be a strong force of resistance if there is a need to mobilize and activate employees to undertake the change (Bernstrøm, 2014). As Clegg and Walsh (2007) have found, employees who undergo a series of changes may not be optimistic or enthusiastic about change because they may feel that it will cost them resources in the job or that their workload will increase. Our experience of almost ten years of research is therefore that the leader plays a key role. Good results are achieved by a good dialogue between the leader and the employees, a dialogue in which they agree on developing the job after a change or an implemented measure. However, there is no reason to think that this will be achieved without any effort. To achieve this goal usually requires a fundamental reorientation of values, strengths, and passions in the organization, among the managers, and within the individual employees.

References

Aagestad, T. S., Tynes, T., Sterud T., Løvseth, K. E., Gravseth, M. H., Eiken, T., … Aasnæss, S. (2011). Faktabok om arbeidsmiljø og helse 2011 [Facts about work environment and health] (STAMI-rapport. Årg. 12 nr. 7). Oslo: Statens arbeidsmiljøinstitutt. Collected from http://www.stami.no/faktabok-om-arbeidsmiljo-og-helse-2011.

Abildgaard, J., Saksvik, P. Ø., & Nielsen, K. (2016). How to measure the intervention process? An assessment of qualitative and quantitative approaches to data collection in the process evaluation of organizational interventions. *Frontiers in Psychology, 7,* 1380. doi:10.3389/fpsyg.2016.01380.

Argyris, C., & Schön, D. A. (1996). *Organizational learning II.* USA: Addison-Wesley Publishing Company.

Armenakis, A. A., & Bedeian, A. G. (1999). Organizational change: A review of theory and research in the 1990s. *Journal of Management, 25,* 293–315.

Arnetz, B. B. (1996). Techno-stress: A prospective psychophysiological study of the impact of a controlled stress-reduction program in advanced telecommunication systems design work. *Journal of Occupational and Environmental Medicine, 38*(1), 53–65.

Bateson, G. (1979). *Mind and nature: A necessary unity (Advances in systems theory, complexity, and the human sciences).* Hampton Press. ISBN 1-57273-434-5.

Bernstrøm, V. H. (2014). Implementering av organisasjonsendringer i helsesektoren—Hvorfor det ofte går galt [Implementation of organizational change in the health sector—Why it often goes wrong]. *Scandinavian Journal of Organizational Psychology, 6*(1), 12–21.

Biron, C., & Saksvik, P. Ø. (2010). Sickness presenteeism and attendance pressure factors: Implications for practice. In C. L. Cooper, J. C. Quick, & M. Schabracq (Eds.), *Work and health psychology. The handbook* (3rd ed.). London: Wiley.

Burke, W. W. (2014). *Organization change: Theory and practice* (4th ed.). Thousand Oaks, CA, London, New Delhi: Sage Publications.

Clegg, C., & Walsh, S. (2007). Change management: Time for a change! *European Journal of Work and Organizational Psychology, 13*(2), 217–239.

Dahl-Jørgensen, C., & Saksvik, P. Ø. (2002). Tiltaket utvidet egenmelding [The measure extended self-employed sickness absenteeism]. In A. Mikkelsen (Eds.), *Tiltak mot sykefravær* [Measures against sickness absenteeism]. Oslo: J.W. Cappelens Forlag.

Dahl-Jørgensen, C., & Saksvik, P. Ø. (2005). An evaluation of the impact of two workplace interventions on the health of service workers. *International Journal of Health Services, 35,* 529–549.

DiFonzo, N., & Bordia, P. (2002). Corporate rumor activity, belief and accuracy. *Public Relations Review, 28*(1), 1–19. doi:10.1016/S0363811102001078.

Dunham, J. (2001). *Stress in the workplace: Past, present and future.* London: Whurr.

Eklöf, M., Ingelgård, A., & Hagberg, M. (2004). Is participative ergonomics associated with better working environment and health? A study among Swedish white-collar VDU users. *International Journal of Industrial Ergonomics, 34*(5), 355–366. doi:10.1016/j.ergon.2004.04.013.

Elo, A. L., Ervasti, J., Kuosma, E., & Mattila, P. (2008). Evaluation of an organizational stress management program in a municipal public works organization. *Journal of Occupational Health Psychology, 13*(1), 10–23.

Finger, M., & Asún, M. (2000). *Adult education at the crossroads. Learning our way out.* London: Zed Books.

Jimmieson, N. L., Terry, D. J., & Callan, V. J. (2004). A longitudinal study of employee adaptation to organizational change: The role of change-related information and change-related self-efficacy. *Journal of Occupational Health Psychology, 9*(1), 11–27. doi:10.1037/1076-8998.9.1.11.

Kelloway, E. K., Hurrell, J. J., Jr., & Day, A. (2008). 20 Workplace interventions for occupational stress. In K. Näswall, J. Hellgren, & M. Sverke (Eds.), *The individual in the changing working life* (pp. 419–441). New York: Cambridge University Press.

Kico, E., & Saksvik. P. Ø. (2015). Derailed, but implemented. In M. Karanika-Murray & C. Biron (Eds.) *Derailed organizational stress and well-being interventions: Confessions of failure and solutions for success.* Springer Education.

Kivimäki, M., Vahtera, J., Pentti, J., Thomson, L., Griffiths, A., & Cox, T. (2001). Downsizing, changes in work, and self-rated health of employees: A 7-year 3-wave panel study. *Anxiety, Stress & Coping: An International Journal, 14*(1), 59–73. doi:10.1080/10615800108248348.

Kritsonis, A. (2005). Comparison of change theories. *International Journal of Scholarly Academic Intellectual Diversity, 8*(1), 1–7.

LaMontagne, A. D., Keegel, T., Louie, A. M., Ostry, A., & Landbergis, P. A. (2007). A systematic review of the job-stress intervention evaluation literature, 1990–2005. *International Journal of Occupational and Environmental Health, 13,* 268–280.

Lewin, K. (1951). *Field theory in social science; selected theoretical papers.* In D. Cartwright (Ed.). New York: Harper & Row.

Lewis, L. K. (2011). *Organizational change: Creating change through strategic communication.* Malaysia: Ho Printing.

Lien, M., & Saksvik P. Ø. (2016). Healthy change processes—A diary study of five organizational units. Establishing a healthy change feedback loop. *Health and Stress.* October edition. doi:10.1002/smi.2698.

Lines, R. (2004). Influence of participation in strategic change: Resistance, organizational commitment and change goal achievement. *Journal of Change Management, 4*(3), 193–215. doi:10.1080/1469701042000221696.

Maurier, W. L., & Northcott, H. C. (2000). Job uncertainty and health status for nurses during restructuring of health care in Alberta. *Western Journal of Nursing Research, 22*(5), 623–641. doi:10.1177/01939450022044638.

Milch, V., Vaag, J., Giæver, F., & Saksvik, P. Ø. (2013). Building healthy organizations through music and culture interventions. In G. Bauer & G. Jenny (Eds.), *Concepts of salutogenic organizations and change: The logics behind organizational health intervention research.* New York: Springer Education.

Murta, S. G., Sanderson, K., & Oldenburg, B. (2007). Process evaluation in occupational stress management programs: A systematic review. *American Journal of Health Promotion, 21*(4), 248–254. doi:10.4278/0890-1171-21.4.248.

Nelson, A., Cooper, C. L., & Jackson, P. R. (1995). Uncertainty amidst change: The impact of privatization on employee job satisfaction and well-being. *Journal of Occupational and Organizational Psychology, 68,* 57–71. doi:10.1111/j.2044-8325.1995.tb00688.x.

Nielsen, K., & Randall, R. (2012). Opening the black box: Presenting a model for evaluating organizational-level interventions. *European Journal of Work and Organizational Psychology, 22*(5), 601–617. doi:10.1080/1359432X.2012.690556.

Nielsen, K., Randall, R., & Albertsen, K. (2007). Participants' appraisals of process issues and the effects of stress management interventions. *Journal of Organizational Behavior, 28*(6), 793–810. doi:10.1002/job.450.

Nguyen, H., & Kleiner, B. H. (2003). The effective management of mergers. *Leadership & Organization Development Journal, 24*(8), 447–454. doi:10.1108/01437730310505876.

Nytrø, K., Saksvik, P. Ø., Mikkelsen, A., Bohle, P., & Quinlan, M. (2000). An appraisal of key factors in the implementation of occupational stress interventions. *Work & Stress, 13,* 213–225.

O'Driscoll, M., & Beehr, T. A. (1994). Supervisor behaviors, role stressors, and uncertainty as predictors of personal outcomes for subordinates. *Journal of Organizational Behavior, 15*(2), 141–155. doi:10.1002/job.4030150204.

Orpen, C. (1994). Interactive effects of work motivation and personal control on employee job performance and satisfaction. *The Journal of Social Psychology, 134*(6), 855–856. doi:10.1080/00224545.1994.9923021.

Orucevic, N. (2013). *Motstand som ressurs?: En longitudinell studie av en omstilling* [Resistance working as a resource?: A longitudinal study of an organizational change] (Unpublished master's thesis). Trondheim: Department of Psychology, NTNU.

Østerud, A. (2014). *En longitudinell studie av endring i Forsvaret* [A longitudinal study of change in the armed forces] (Unpublished master's thesis). Trondheim: Department of Psychology, NTNU.

Øyum, L., Andersen, T. K., & Saksvik, P. Ø. (2006). *Good leadership in healthy change processes —Case studies on how managers have made the change process a positive experience for the employees*. Paper presented at the EA-OHP Annual Conference, Dublin, Ireland.

Paulsen, N., Callan, V. J., Grice, T. A., Rooney, D., Gallois, C., Jones, E., et al. (2005). Job uncertainty and personal control during downsizing: A comparison of survivors and victims. *Human Relations, 58,* 463–495. doi:10.1177/0018726705055033.

Peters, T. J., & Waterman, R. H. (1982). *In search of excellence*. NY: Warner.

Pollard, T. M. (2001). Changes in mental well-being, blood pressure and total cholesterol levels during workplace reorganization: The impact of uncertainty. *Work & Stress, 15*(1), 14–28. doi:10.1080/02678370110064609.

Publicom Forlag. (2014). http://www.publicom.no/geir-lippestad-foredrag/?gclid=CMPvuZPEk8ICFQLecgod9aQAyg.

Rafferty, A. E., & Griffin, M. A. (2006). Perceptions of organizational change: A stress and coping perspective. *Journal of Applied Psychology, 91,* 1154–1162. doi:10.1037/0021-9010.91.5.1154.

Randall, R., Griffiths, A., & Cox, T. (2005). Evaluating organizational stress-management interventions using adapted study designs. *European Journal of Work and Organizational Psychology, 14*(1), 23–41. doi:10.1080/13594320444000209.

Randall, R., Nielsen, K. M., & Tvedt, S. D. (2009). The development of five scales to measure employees' appraisals of organizational-level stress management interventions. *Work & Stress, 23,* 01–23.

Reynolds, S. (1997). Psychological well-being at work: Is prevention better than cure? *Journal of Psychosomatic Research, 43*(1), 93–102. doi:10.1016/s0022-3999(97)00023-8.

Richardson, K. M., & Rothstein, H. R. (2008). Effects of occupational stress management intervention programs: A meta-analysis. *Journal of Occupational Health Psychology, 13*(1), 69–93. doi:10.1037/1076-8998.13.1.69.

Saksvik, P. Ø. (2012). *Hunting the trolls of interventions.* In Keynote presentation at the IPEP-Conference, Nottingham 23–24 May.

Saksvik, P. Ø., & Nytrø, K. (2001). Improving subjective health and reducing absenteeism in a natural work life-intervention. *Scandinavian Journal of Psychology, 42,* 17–24.

Saksvik, P. Ø., & Nytrø, K. (2005). Hvordan virksomheter kan forebygge og håndtere belastninger på arbeidsplassen [On how Enterprises can prevent and handle stress at work]. In S. Einarsen & A. Skogstad (Eds.), *Den dyktige medarbeider [The skilled employee]* (pp. 391–408). Bergen: Fagbokforlaget.

Saksvik, P. Ø., & Nytrø, K. (2011). Utfordringer med å drive systematisk helse-, miljø- og sikkerhetsarbeid i et deregulert arbeidsliv. In Saksvik, P. Ø. (ed.). *Arbeids- og organisasjonspsykologi – aktuelle tema til inspirasjon for et bedre arbeidsliv*. Oslo. Cappelen Akademisk Forlag.

Saksvik, P. Ø., Nytrø, K., Dahl-Jørgensen, C., & Mikkelsen, A. (2002). A process evaluation of individual and organizational occupational stress and health interventions. *Work & Stress, 16,* 37–57.

Saksvik, P. Ø., Olaniyan, S. O., Lysklett, K., Lien, M., & Bjerke, L. (2015). A process evaluation of a salutogenic intervention. *Scandinavian Psychologist, 2,* e8. http://dx.doi.org/10.15714/scandpsychol.2.e8.

Saksvik, P. Ø., & Tvedt, S. D. (2009). Leading change in a healthy way. *Scandinavian journal of Organizational Psychology, 1,* 20–29.

Saksvik, P. Ø., Tvedt, S. D., Nytrø, K., Buvik, M. P., Andersen, G. R., Andersen, T. K., et al. (2007). Developing criteria for healthy organizational change. *Work & Stress., 21,* 243–263.

Tims, M., & Bakker, A. B. (2010). Job crafting: Towards a new model of individual job redesign. *South African Journal of Industrial Psychology, 36*, 1–9.

Tosey, P., Visser, M., & Saunders, M. N. K. (2012). The origins and conceptualizations of 'triple-loop' learning: A critical review. *Management Learning, 43*(3), 291–307.

Tvedt, S. D., & Saksvik, P. Ø. (2012). Finding the content of the process part of interventions. In C. Biron, M. Karanika-Murray, & C. L. Cooper (Eds.), *Managing psychosocial risks in the workplace: The role of process issues*. Routledge/Psychology Press.

Tvedt, S. D., Saksvik, P. Ø., & Nytrø, K. (2009). Healthy organizational change processes as a prerequisite for a good psychological work environment. *Work & Stress, 23*, 80–98.

Vaag J., Saksvik, P. Ø., Theorell, T., Skillingstad, T., & Bjerkeset, O. (2012) Sound of well-being—Choir singing as an intervention to improve well-being among employees in two Norwegian county hospitals, *Arts & Health: An International Journal for Research, Policy and Practice, iFirst article,* 1–10. doi:10.1080/17533015.2012.727838.

Visser, M. (2002). *Managing knowledge and action in organizations; towards a behavioral theory of organizational learning*. In EURAM Conference, Organizational Learning and Knowledge Management, Stockholm, Sweden.

Wilson, M. G., Basta, T. B., Bynum, B. H., DeJoy, D. M., Vandenberg, R. J., & Dishman, R. K. (2010). Do intervention fidelity and dose influence outcomes? Results from the move to improve worksite physical activity program. *Health Education Research, 25*(2), 294–305. doi:10.1093/her/cyn065.

Wrzesniewski, A., & Dutton, J. E. (2001). Crafting a Job: Revisioning employees as active crafters of their work. *The Academy of Management Review, 26*(2), 179–201.

Chapter 13
Healthy Individuals in Healthy Organizations: The Happy Productive Worker Hypothesis

Marit Christensen

The individual is the primary constituent of and the driving force behind an organization. Sickness, health-related issues, presenteeism, and sickness absenteeism could become a threat for the organization in terms of increased costs and production loss (Boles, Pelletier, & Lynch, 2004; Burton et al., 2005). On this basis, occupational health has focused on the negative aspects of working life (Christensen, Aronsson, Clausen, Hakanen, & Vivoll Straume, 2012; Halbesleben & Buckley, 2004; Wright & Cropanzano, 2004). Over time, this negative focus has been more important than the focus on health promotion and the positive factors in the working life. This is not necessarily always appropriate since workers who do well both mentally and physically also appear to be more healthy and productive (Fisher, 2003). Based on this, we will take a closer look at which factors are preventive and which are promotional for the workers experience of the psychosocial working environment, as well as at the connection between positive factors in the working life and productivity.

Today, working life in Norway is distinguished by a number of challenges, something that is unique in Norway and some that we share with other parts of the world. The aging of the population is rapidly accelerating (White Paper 29, 2012–2013) and will lead to great costs for both the employers and the Norwegian state and, in some sectors, will bring about a lack of manpower. There will be a need for people to work longer while still experiencing job satisfaction and being healthy and productive. In other words, we need an inclusive working life. One of the possible solutions to the need may be to build attractive and resourceful jobs, while making the workers resilient, adaptable, and able to manage stress in a world of global competition, economic crises, increasing demands, constant change, and ongoing pressure to produce. From this perspective, it is important to look at what

M. Christensen (✉)
Department of Psychology, Norwegian University of Science and Technology (NTNU), Trondheim, Norway
e-mail: marit.christensen@ntnu.no

M. Christensen et al. (eds.), *The Positive Side of Occupational Health Psychology*,
https://doi.org/10.1007/978-3-319-66781-2_13

prevents negative health-related consequences for employees, what promotes well-being and health, and, last but not least, what creates productive and sustainable organizations. Among others, positive occupational health psychology offers a focus that can contribute to managing constant change and the increasing demands in the working environment. Research has shown that positive affectivity has a positive correlation with creativity (Isen, Daubman, & Nowicki, 1987), resilience or resistance (Fredrickson, Tugade, Waugh, & Larkin, 2003), and the ability to manage stress (Glazer & Kruse, 2008), as well as to a reduction in the risks of long-term sickness absenteeism (Clausen, Christensen, & Borg, 2010). It also increases the possibility of returning to work after long-term sick leave (Clausen, Friis Andersen, Bang Christensen, & Lund, 2011). For example, Harter, Schmidt, and Hayes (2002) showed that work engagement is positively correlated with a number of positive organizational outcomes. Work engagement appears to be connected to production-related outcomes such as job performance (Bakker & Bal, 2010; Halbesleben & Wheeler, 2008), satisfied customers (Salanova, Agut, & Peiró, 2005), and financial profit (Xanthopoulou, Bakker, Demerouti, & Schaufeli, 2009).

Based on Kivimäki and Lindström (2006), a healthy organization has a two-sided focus in which the employee's well-being and the organization's performance are simultaneously in focus. Whether an organization is healthy depends on the "the capability of the organization to function effectively in relation to various environmental factors and to respond to various environmental changes" (Kivimäki & Lindström, 2006, p. 811). The term healthy organizations constitutes a focal point that allows concepts such as working environment, health, and productivity to coexist in achieving the optimal level where well-being at an individual level exists side by side with effective and productive organizations (Christensen et al., 2012).

A research group in Spain has developed the concept of "HERO" (healthy and resilient organizations). The HERO model considers the interaction between individuals who work together and offers a macro-view that makes it possible to integrate the different levels of analysis (organization, group, and individual) and the study of organizational phenomena (Salanova, Martínez, & Llorens, 2014). In this case, healthy means: (a) Organizations are focused on caring for both the employee's health and the organization as a whole (this includes its effectiveness, survival, and future development); and (b) it is possible to differentiate between healthy and unhealthy organizations, something this indicates that there are methods for structuring and for leading the working process in a healthier direction (Salanova, Llorens, Cifre, & Martínez, 2012). Organizations are referred to as resilient because they adapt positively to challenging circumstances, become stronger in unfortunate situations, and maintain their functions and results in pressured situations (Salanova et al., 2014). A perception that has been strong in both practice and in research is that happy workers are productive workers [the happy productive worker hypothesis (Wright, Cropanzano, Denney, & Moline, 2002)]. The big question is whether the research actually supports this hypothesis. It becomes reasonable to ask: What is a healthy individual in a healthy organization?

To be able to answer this, it is worth taking a closer look at what characterizes a happy and healthy employee, and what defines a healthy organization.

In this chapter, we wish to present a critical view of what lies beyond the mystery of "the happy productive worker hypothesis" and, thus, to perhaps come closer to the answer to what gives rise to healthy individuals in healthy organizations.

13.1 The Happy Productive Worker Hypothesis

"The happy productive worker hypothesis" has a long history and has been called the Holy Grail in occupational theory (Landy, 1985). Research on the hypothesis had already begun in the 1930s with the human relation movement (Wright, Cropanzano, & Bonett, 2007), and research related to the hypothesis is still very popular today. Even after so many years, a lot of skepticism is associated with the correlation between subjective well-being and high productivity (Wright & Staw, 1999). Many of the reasons for this are associated with the fact that meta-analyses have found weak positive correlations between subjective well-being and productivity (Wright & Cropanzano, 2000). The strength of the correlation between subjective well-being and productivity is different in different studies, something that indicates that the way the concepts are operationalized explains some of the equivocal findings. The unclear correlation can be explained by a different operationalization of both subjective well-being and productivity (Wright & Cropanzano, 2000). Most of the studies that are conducted have operationalized work-related subjective well-being as job satisfaction (Wright et al., 2002).

13.2 Conceptualization of Happiness

The understanding of what the term subjective well-being consists of is essential for understanding the relationship between subjective well-being and productivity. How does the term define itself, and what does it consist of? Subjective well-being refers to how individuals evaluate their lives (Diener, Sandvik, & Pavot, 1991). The evaluation can either happen through cognition on how an individual evaluates their satisfaction with life in general, or it can be in the form of affectivity, such as feelings of comfort or discomfort toward their life (Diener et al., 1991). In his circumplex model (see Fig. 10.1), Russell (1980, 2003) suggests that affective states occur in two fundamental neurophysiologic systems. The first is related to the happiness continuum, and the second is related to the activation continuum. Different variants of subjective well-being (work engagement, job satisfaction, workaholism, and burnout) can be understood as a linear combination of these two dimensions in a varied degree of both pleasure and activation. As one can see from the model, the degree of activation varies when one experiences the different variants of subjective well-being (Warr, 2007). For example, when one is pleased and relaxed this means that there is a lower level of activation than when one is enthusiastic and energetic. According to Diener et al.'s (1991) definition of subjective well-being, an employee has a high subjective well-being when he or she experiences being pleased with his or her job, often experiencing positive emotions and seldom experiencing negative emotions. The employees can either experience a high activation in the job (engagement, workaholism) or a low activation (job satisfaction, burnout).

When subjective well-being is operationalized as a more cognitive-oriented concept, such as satisfaction, the correlation with productivity becomes weaker than when subjective well-being is measured as an affective condition, such as work engagement (Taris & Schreurs, 2009; Wright & Cropanzano, 2000). An explanation for this can be that cognitive-related concepts such as job satisfaction are often defined as more passive attitudes to the job, whereas well-being-related concepts, such as engagement and burnout, have a more active component (Reijseger, Schaufeli, Peeters, & Taris, 2013). On the one hand, job satisfaction concerns the employee's evaluation of their job, whereas engagement is a motivational condition (Schaufeli & Bakker, 2004; Wright & Cropanzano, 2004). This is illustrated in Fig. 13.1, which shows a circumplex model for affect related to work-related conditions (Bakker & Oerlemans, 2011a; Russell & Carroll, 1999; Russell, 1980, 2003). It will be useful at this point to take a closer look at the concepts.

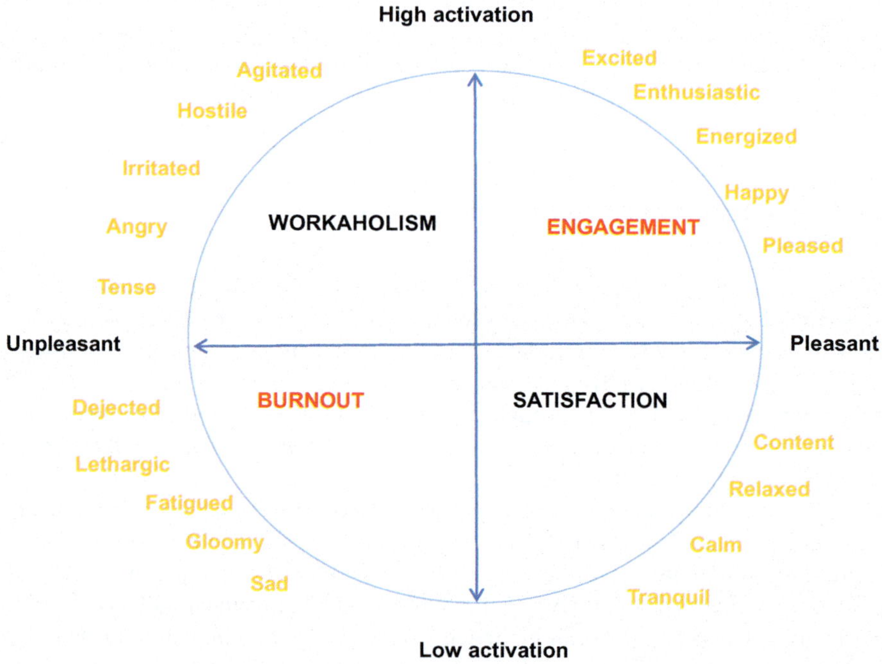

Fig. 13.1 Circumplex model of job-related subjective well-being (Bakker & Oerlemans, 2011a; Russell, 1980, 2003; Russell & Carroll, 1999). *Source* Bakker and Oerlemans (2011b)

13.2.1 Work Engagement

Based on the classification system, work engagement is characterized by a high level of happiness and a high level of activation. For the most part, work engagement is defined as "… a positive, fulfilling, affective-motivational work-related state of mind that is characterized by vigor, dedication and absorption" (Schaufeli, Salanova, González-Romá, & Bakker, 2002, p. 74; Schaufeli & Bakker, 2010). Vigor is characterized by a high level of energy and mental endurance when working, a willingness to invest effort in the job, and endurance when faced with challenges. Dedication refers to being strongly involved in the job, identifying oneself with it, experiencing it as important, and as being inspired and proud of one's job and viewing it as a challenge. Absorption is characterized by being deeply concentrated and totally absorbed by work tasks such that time flies and one finds it difficult to detach oneself from the work. The definition of vigor describes a continuing affective-cognitive condition rather than a short-term specific emotional condition. Work engagement is the part of the motivational process where job resources contribute to creating activity, effectiveness, and performance in the job (Bakker, Demerouti, Taris, Schaufeli, & Schreurs, 2003).

Engaged employees are characterized by high energy and enthusiasm that is related to work tasks (Bakker, Schaufeli, Leiter, & Taris, 2008). Reijseger et al. (2013) offer an explanation for the correlation between work engagement and productivity. Their explanation is that the affective-motivational condition entailed in work engagement opens up the employee's perceptions of more opportunities in their environment because positive affectivity expands the cognitive aspects (Fredrickson, 2001). This is based on the "b-and-build theory," which postulates that positive emotions expand the human's immediate thought—and action—repertory. For example, happiness at work can promote a desire to play around and be more creative, and an interest can promote a desire to explore and to learn more (Fazio, Eiser, & Shook, 2004). These expanding thought patterns are often contrasted with thought patterns that are promoted by negative emotions and have a restrictive effect. For example, if someone is feeling frightened or angry, the only reactions are to either get away quickly or to attack. Furthermore, Fredrickson (2001) argued that the consequences of these expanded thought patterns, demonstrated, for example, through play and exploration, will contribute to innovative thinking and creative ideas, as well as to building social networks. This will in turn contribute to building the employee's resources, including their physical, intellectual, social, and psychological resources. Expanding one's thinking repertory through positive emotions will lead to the acquisition of new knowledge and new abilities (Fredrickson & Losada, 2005).

Most of the studies that have looked at the correlation between work engagement and productivity have been based on the subjective reporting of one's own job execution. However, there are also studies that have examined the relationship between work engagement and more objective measures of productivity. Salanova et al. (2005) completed a study of a group of hotel and restaurant employees in Spain in which they examined the relationship between organizational resources, engagement, service climate, the customer's evaluation of the employees' performance, and customer loyalty. The results showed that organizational resources and work engagement predicted a service climate that, in turn, predicted the employee's performance and customer loyalty. Here, the customers evaluated the employee's job performance and customer loyalty was measured by the likelihood of them returning to the establishment. Another example is a study conducted by Xanthopoulou, Bakker, Demerouti, and Schaufeli (2012) who examined how daily variations in job resources are related to personal resources, work engagement, and financial returns. Their sample population consisted of 42 employees in three different businesses in the fast-food industry in Greece. The results show that strengthening job resources, and especially by introducing coaching, can create engaged and productive employees. In this study, productivity was also measured objectively. Financial returns referred to the total amount earned by a specific shift, and these amounts were matched with the records of the relevant shift. Hernández-Vargas, Llorens-Gumbau, and Rodríguez-Sánches (2014) examined a private hospital in Spain. The results showed that positive affectivity and work engagement mediate the relationship between the employee's self-efficacy and the quality of service. Employees who had a great deal of self-efficacy demonstrated high levels of positive affectivity and a greater work engagement, which again lead to greater quality of the service they provided.

Another concept that measures subjective well-being at work is that of job satisfaction, and based on the circumplex model shown in Fig. 10.1, it is the active dimension that differentiates work engagement and job satisfaction.

13.2.2 Job Satisfaction

Job satisfaction has been defined differently, and researchers do not agree on how best to define the term. Some suggest it concerns how satisfied an employee is with his or her job, whether the employee likes the job or some of the different aspects of it (Spector, 1997). Others view the definition of job satisfaction as much more complicated and believe that multidimensional psychological responses to the job are involved (Hulin & Judge, 2003). A definition used by Hulin and Judge (2003) highlights that job satisfaction contains multidimensional psychological responses to one's job and, furthermore, that these personal responses have cognitive, affective, and behavioral components. One important thing to note is that there is considerable variation on whether one defines job satisfaction as feelings about the job (affective job satisfaction) or as cognitions about the job (cognitive job satisfaction) (Moorman, 1993; Thompson & Phua, 2012). Affective job satisfaction is a subjective concept that represents a feeling the employees hold about their jobs

(Spector, 1997). Cognitive job satisfaction is a more objective and logical evaluation of different aspects of the job. These two concepts are different and have different predictors and consequences (Spector, 1997; Thompson & Phua, 2012). The different measurements of job satisfaction can also have different contexts in terms of productivity (Schleicher, Watt, & Greguras, 2004). Following a review of 74 studies, early meta-analyses show that different parts of the job have a weak correlation (0.17) with productivity (Iaffaldano & Muchinsky, 1985). A meta-analysis completed by Judge, Thoresen, Bono, and Patton (2001) showed that general job satisfaction with an affective content has a moderate strong correlation (0.30) with productivity. These meta-analyses indicate that general job satisfaction has a stronger correlation with job performance, compared to job satisfaction with certain aspects of a cognitive incline. The context here is, of course, dependent on how productivity is measured. If it is measured using self-reports of job performance, then this affects the strength of the context since affectivity will affect the employee's reflection of his or her own efforts. Several of the studies in these meta-analyses are cross-sectional, meaning that it is difficult to determine the causal relationship. Job satisfaction can create productivity, but can also be a cause of high productivity (Lyubomirsky, King, & Diener, 2005). There exist longitudinal studies on this subject. Research by Lyubomirsky et al. (2005) that analyzed ten studies found that happiness at work had a positive effect on the employee's job performance. Furthermore, in the classification model, there are two concepts that represent the negative side of subjective well-being: burnout and workaholism. We will take a closer look at these concepts and their relation to productivity below.

13.2.3 Burnout

Burnout is classified with lower levels of both pleasure and activation. Burnout is known as a job-related condition (Maslach & Jackson, 1984). Maslach (1982) defines burnout as a syndrome that is consistent with emotional exhaustion, cynicism/depersonalization, and reduced experience of job performance that can arise in individuals who work with humans (Halbesleben & Buckley, 2004). Burnout can be viewed as a psychological syndrome that arises when employees are exposed to stressful working environments with high job demands and few resources (Demerouti & Bakker, 2008; Maslach, Schaufeli, & Leiter, 2001). Maslach et al.'s three-dimensional definition is without a doubt the definition of burnout that has been researched the most, and has received the most support (Maslach et al., 2001). The term burnout did not originally have a standard definition, although there was consensus on the three core components of burnout. Continued research on this field led to the development of a multidimensional theory of burnout, and this theoretical framework continues to be dominant within this field.

Exhaustion is viewed as the most central and obvious feature of burnout. Of the three dimensions, this component has been the most reported and analyzed.

Exhaustion reflects the stress dimension of burnout, promotes behavior that distances one emotionally and cognitively from the job, and refers to a feeling of having exhausted one's emotional and physical resources. Depersonalization/cynicism can be viewed as behavior that creates a distance between oneself and the service recipients by actively ignoring the quality that makes them engaged people. This condition of distance can be viewed as an immediate reaction to burnout. This component refers to a negative and insensitive, or a comprehensive, distance in responses toward varied aspects of the job. Depersonalization refers mostly to occupations that have clients, whereas cynicism is more likely to involve negative emotions/indifference toward the job. Feelings of ineffectiveness or a reduced experience in job performance can come from a working situation with chronic overwhelming demands that contribute to exhaustion and cynicism. This dimension represents a self-evaluating dimension of burnout and refers to feelings of incompetence and reduced experience of job performance and productivity in the job. There is a great deal of discussion in the literature about whether this dimension is actually part of the construct. The lack of self-efficacy seems to arise from the lack of relevant resources, whereas exhaustion and cynicism seem to arise from job demands that include an overwhelming workload and social conflict (Maslach et al., 2001).

Research on the relation between burnout and productivity is unclear (Bakker, Demerouti, & Verbeke, 2004). Some of this lack of clarity may be the result of the lack of a clear theoretical background. Employees who are emotionally exhausted can sense that they lack adaptive resources, and that they lack the ability to give more to their job. The energy they once possessed is gone, and they are now left without the resources to perform adequately. Depersonalization, also known in the literature as cynicism and disengagement, arises as a response to emotional exhaustion and describes a process where the employees feel distanced from their job and where they create cold or indifferent attitudes toward their job performance and their clients (and colleagues). Reduction in experienced job performance refers to a sense of weakened ability to perform in the job. Employees that experience burnout are more dissatisfied with their jobs, less attached to the organization, more distant, and perform more poorly in the job. The increasing amount of long-term sick leave due to burnout causes a significant burden for the employees, the employer, and the insurance companies (Maslach et al., 2001; Peterson et al., 2008). In their study, Bakker et al. (2004) used the JD-R model as a theoretical basis and found clear correlations, even though the correlations were not strong. One explanatory factor for the low correlation may be the operationalization of performance and productivity measures. The findings indicate that future research should take a closer look at how these are measured (Bakker et al., 2004).

13.2.4 Workaholism

Workaholism is classified with a low level of pleasure and a high level of activation. Workaholism is defined by a strong inner driving force to work excessively hard (Oates, 1971; Schaufeli, Taris, & Van Rhenen, 2008b). Workaholic employees have a compulsive driving force to working continually and to use a disproportionate amount of time on the job; they are addicted to the job (Clark, Michel, Zhdanova, Pui, & Baltes, 2014) and therefore work more than is expected of the organizational demands (Taris, Schaufeli, & Shimazu, 2010). Workaholics work so much that they have few resources left for family life or for others activities outside of job-related activities (Bonebright, Clay, & Ankenmann, 2000). The obsessional, joyless force is what differentiates workaholics from, for example, work engagement, where joy while working is a large part of the definition (Schaufeli, Taris, & Bakker, 2008a). There is consensus in the research that some affective experiences, such as anxiety and guilt, are expressed when one is not working (Ng, Sorensen, & Feldman, 2007; Spence & Robbins, 1992) but one is more uncertain about what is going on when actually at work. Some have proposed that workaholism is characterized by low work enjoyment (Spence & Robbins, 1992), whereas others conclude that workaholics enjoy working (Ng et al., 2007).

Clark et al. (2014) conducted a meta-analysis on workaholism, and found that workaholics engaged in a great deal of negative behavior, became more stressed and had poorer health. In addition, in the long run, these employees ended up being less productive than their colleagues. Researcher has also shown that workaholics have a tendency to have more negative interactions with their coworkers (Ng et al., 2007). They have the tendency to be perfectionists, have unrealistically high standards, and their behavior can also foster competition with the coworkers. These behavioral patterns have a negative effect on productivity. The connection between work engagement and workaholism has received mixed support. Workaholism seems to be positively correlated with general work engagement and absorption, whereas no connection has been found between workaholism and the dimensions of vigor and dedication. This may be consistent with the idea that workaholics experience some kind of "rush" from working (Sussman, 2012), a phenomenon that resembles the absorption dimension in work engagement where employees are totally absorbed and have difficulty in disengaging themselves (McMillan, O'Driscoll, & Burke, 2003). A meta-analysis concerning the relationship between work engagement and a number of overlapping concepts showed, on the one hand, a positive connection between work engagement and organizational attachment, as well as work engagement and job involvement. On the other hand, the results showed a negative connection between workaholism and positive affectivity, job performance, and job satisfaction (Christian, Garza, & Slaughter, 2011). Here, one may conclude that workaholism and work engagement share some common characteristics, namely being involved and absorbed in one's work. In their study, Clark et al. (2014) found a weak-to-moderate correlation between workaholism and happiness at work, and a positive connection with negative affectivity. The authors suggest that a possible explanation for this is that the employees' negative emotions will make up for the positive feelings workaholics have when they work. The study also showed that workaholics are more likely to experience as much negativity at work as they do at home (Clark et al., 2014). Yet, although workaholics experience positive feelings and a rush from working, this seems to be short-lasting and something that is easily replaced by negative emotions.

The results of the meta-analysis clearly showed that workaholism was related to negative individual and organizational variables (Clark et al., 2014). Workaholism was not related to high levels of performance or to job satisfaction, but rather to negative outcomes such as burnout, job stress, low job satisfaction, poor mental, and poor physical well-being. Workaholics do not seem to be productive employees, and they appear to cost the organization more in the long term because of their poor health and well-being (Clark et al., 2014; Ng et al., 2007). From the reviews, one can see that the connection between the four categories of subjective well-being and productivity is dependent not only on how one defines and understands subjective well-being, but also on how one defines productivity (Clark et al., 2014; Ng et al., 2007).

13.3 Conceptualization of Productivity

Job performance or productivity has been operationalized and measured in many different ways and on many different levels (individual, team, and organizational levels). This has contributed to an uncertainty around equivocal findings on the connection between subjective well-being and productivity. Job performance has been measured subjectively, where the employees evaluate their own job situation. Furthermore, the employee's job performance has been evaluated and measured by other people, for example, by their managers, customers, users, and patients. There also exist a number of objective measures of job performance, such as registered data on sick leave and financial returns in organizations. These different definitions, operationalization, and measures do not exclude each other; rather, they are complementary (Fay & Sonnentag, 2010). In the definitions and the operationalization, performance indicates both behavior and outcome. These can be divided into process performance and outcome performance (Roe, 1999). Process performance includes behavior or attitudes that the employees use to obtain performance or an effective outcome. Outcome performance is the product in itself or the service that is being performed. The process comes before the outcome, and most of the studies include one of the two, rather than a combination of both. Many studies have shown the correlation between subjective well-being and self-reported job performance, whereas there are fewer studies that show support for the correlation between subjective well-being and objective measures of productivity (Demerouti & Cropanzano, 2010). Precisely because of the multidimensional nature of the term, subjective well-being will affect the separate dimensions differently.

13.4 Implications and Conclusion

The clarification of concepts is the key for getting to the bottom of understanding "the happy productive worker hypothesis." We have to understand what it is we are researching and how we can operationalize and measure it in the correct way. This includes both the concept of subjective well-being and that of productivity. This is important for research, and not at least for practice, in order to ensure that we use the appropriate measure based on an evidence-based understanding of creating subjective well-being and, at the same time, achieve the expected increase in productivity.

The concepts that have been discussed in this chapter are to some extent overlapping, and research has not yet clarified a cut-off point between, for example, engagement and workaholism. Can work engagement become something negative and if so, when does it become negative? Jobs that create a great deal of work engagement are often challenging jobs that on the one hand offer personal growth, but on the other hand can cause stress, overload, and burnout. When one is engaged, one becomes absorbed and forgets time and place, something that may affect the work–family balance because the individual spends more time on the job than with the family. Engaged people work because it brings them happiness. In contrast, workaholics appear to have an obsessive drive to work and experience negative emotions when they do not work. A finding that balances this out shows that people with high work engagement are also often engaged in activities outside of work (Gorgievski, Bakker, & Schaufeli, 2010).

In their article, Salanova, Llorens, Acosta, and Torrente (2013) discussed the question of what creates healthy and resilient organizations (HERO) and they raised several points. When one maps and evaluates organizations, it is advantageous to match the right people and teams to the correct organization and to create an optimal balance between the team and the organization's values and goals. Furthermore, it is important to design and change the workplace and organizational practice in order to reduce psychological risk factors and promote healthy employees and healthy organizations. A further point they mention is the need to promote transformational leadership, which they view as a key resource for creating health and subjective well-being. Exercises on promoting self-efficacy are also highlighted as important, and the development career planning that could be a key practice for developing healthy and productive organizations, especially in situations of job insecurity. The survey instrument, Nordic Questionnaire on Positive Organizational Psychology (N-POP), has been developed on the basis of mapping positive factors in the working life in order to obtain an understanding of the employee's experience of their job-related conditions and the individual and organizational outcomes (Christensen et al., 2012). This instrument can be a useful addition for identifying both existing and potential resources in the working environment and, thereafter, for strengthening them to create a healthy organization. This can contribute to building positive qualities such as personal growth, work engagement, and healthy practices in an organization. It may also be useful for building a base for interventions, document changes, and for the evaluation of organizational interventions, as well as for supporting.

References

Bakker, A. B., & Bal, M. P. (2010). Weekly work engagement and performance: A study among starting teachers. *Journal of Occupational and Organizational Psychology, 83*(1), 189–206. doi:10.1348/096317909X402596.

Bakker, A. B., Demerouti, E., Taris, T., Schaufeli, W. B., & Schreurs, P. (2003). A multi-group analysis of the job demands-resources model in four home care organizations. *International Journal of Stress Management, 10,* 16–38.

Bakker, A. B., Demerouti, E., & Verbeke, W. (2004). Using the job demands-resources model to predict burnout and performance. *Human Resource Management, 43*(1), 83–104.

Bakker, A. B., Oerlemans, W. G. M. (2011a). Subjective well-being in organizations. In K. S. Cameron & G. M. Spreitzer (Red.), *Handbook of positive organizational scholarship.* Oxford University Press.

Bakker, A., Oerlemans, W. (2011b). A two-dimensional view of well-being. In K. S. Cameron & G. M. Spreitzer (Eds.), *The Oxford handbook of positive organizational scholarship*. Oxford University Press. doi:10.1093/oxfordhb/9780199734610.001.0001.

Bakker, A. B., Schaufeli, W. B., Leiter, M. P., & Taris, T. W. (2008). Work engagement: An emerging concept in occupational health psychology. *Work & Stress, 22,* 187–200.

Boles, M., Pelletier, B., & Lynch, W. (2004). The relationship between health risks and work productivity. *Journal of Occupational and Environmental Medicine, 46*(7), 737–745.

Bonebright, C. A., Clay, D. L., & Ankenmann, R. D. (2000). The relationship of workaholism with work-life conflict, life satisfaction, and purpose in life. *Journal of Counseling Psychology, 47*(4), 469–477. doi:10.1037/0022-0167.47.4.469.

Burton, W. N., Chen, C.-Y., Conti, D. J., Schultz, A. B., Pransky, G., & Edington, D. W. (2005). The association of health risks with on-the-job productivity. *Journal of Occupational and Environmental Medicine, 47*(8), 769–777.

Christensen, M., Aronsson, G., Clausen, T., Hakanen, J., & Vivoll Straume, L. (2012). *Building engagement and healthy organisations. A test of the Nordic questionnaire on Positive Organisational psychology (N-POP).* TemaNord 2012: 549.

Christian, M. S., Garza, A. S., & Slaughter, J. E. (2011). Work engagement: A quantitative review and test of its relations with task and contextual performance. *Personnel Psychology, 64,* 89–136. doi:10.1111/j.1744-6570.2010.01203.x.

Clark, M. A., Michel, J. S., Zhdanova, L., Pui, S. Y., & Baltes, B. B. (2014). All work and no play? A meta-analytic examination of the correlates and outcomes of workaholism. *Journal of Management, Advance online publication.* doi:10.1177/0149206314522301.

Clausen, T., Christensen, K. B., & Borg, V. (2010). Positive work-related states and long-term sickness absence: A study of register-based outcomes. *Scandinavian Journal of Public Health, 38*(Suppl 3), 51–58.

Clausen, T., Friis Andersen, M., Bang Christensen, K., & Lund, T. (2011). Return to work among long-term sickness absent employees in eldercare—A prospective analysis of register-based outcomes. *International Journal of Rehabilitation Research, 34*(3), 249–254.

Demerouti, E., & Bakker, A. B. (2008). The Oldenburg Burnout Inventory: A good alternative to measure burnout and engagement. In J. R. B. Halbesleben (Ed.), *Handbook of stress and burnout in health care*. Nova Science: Hauppauge, NY.

Demerouti, E., & Cropanzano, R. (2010). From thought to action: Employee work engagement and job performance. In A. B. Bakker & M. P. Leiter (Eds.), *Work engagement: A handbook of essential theory and research* (pp. 147–163). New York: Psychology Press.

Diener, E., Sandvik, E., & Pavot, W. (1991). Happiness is the frequency, not the intensity, of positive versus negative affect. In F. Strack, M. Argyle, & N. Schwarz (Eds.), *Subjective well-being: An interdisciplinary perspective*. New York: Pergamon.

Fay, D., & Sonnentag, S. (2010). A look back to move forward: New directions for research on proactive performance and other discretionary work behaviors. *Applied Psychology: An International Review, 59,* 1–20.

Fazio, R. H., Eiser, J. R., & Shook, N. J. (2004). Attitude formation through exploration: Valence asymmetries. *Journal of Personality and Social Psychology, 87,* 293–311.

Fisher, C. D. (2003). Why do lay people believe that job satisfaction and job performance are related? Possible sources of a common sense theory. *Journal of Organizational Behavior, 24,* 753–777.

Fredrickson, B. L. (2001). The role of positive emotions in positive psychology: The broaden and build theory of positive emotions. *American Psychologist, 56,* 218–226.

Fredrickson, L., & Losada, M. F. (2005). Positive affect and the complex dynamics of human flourishing. *American Psychologist, 68*(9), 822. doi:10.1037/a0034435.

Fredrickson, B. L., Tugade, M. M., Waugh, C. E., & Larkin, G. R. (2003). What are good emotions in crisis? A prospective study of resilience and emotions following the terrorist attacks on the United States on September 11th, 2001. *Journal of Personality and Social Psychology, 84*(2), 365–376.

Glazer, S., & Kruse, B. (2008). The role of organizational commitment in occupational stress models. *International Journal of Stress Management, 15,* 329–344.

Gorgievski, M., Bakker, A. B., & Schaufeli, W. B. (2010). Work engagement and workaholism: Comparing the self-employed and salaried employees. *Journal of Positive Psychology, 5,* 83–96.

Halbesleben, J. R. B., & Buckley, M. R. (2004). Burnout in organizational life. *Journal of Management, 30*(6), 859–879.

Halbesleben, J. R. B., & Wheeler, A. R. (2008). The relative roles of engagement and embeddedness in predicting job performance and intention to leave. *Work and Stress, 22,* 242–256.

Harter, J. K., Schmidt, F. L., & Hayes, T. L. (2002). Business unit-level relationship between employee job performance and work force productivity. *American Psychologist, 38,* 473–478.

Hernández-Vargas, C. I., Llorens-Gumbau, S., & Rodríguez-Sánches., A. M. (2014). Healthy employees and service quality in the healthcare sector. *Anales de psicología, 30*(1), 247–258.

Hulin, C. L., & Judge, T. A. (2003). Job attitudes. In W. C. Borman, D. R. Ligen, & R. J. Klimoski (Eds.), *Handbook of psychology: Industrial and organizational psychology* (pp. 255–276). Hoboken, NJ: Wiley.

Iaffaldano, T., & Muchinsky, P. M. (1985). Job satisfaction and job performance: A meta-analysis. *Psychological Bulletin, 97*(2), 251–273. doi:10.1037/0033-2909.97.2.251.

Isen, A. M., Daubman, K. A., & Nowicki, G. P. (1987). Positive affect facilitates creative problem solving. *Journal of Personality and Social Psychology, 52,* 1122–1131.

Judge, T. A., Thoresen, C. J., Bono, J. E., & Patton, G. K. (2001). The job satisfaction–job performance relationship: A qualitative and quantitative review. *Psychological Bulletin, 127,* 376–407.

Kivimäki, M., & Lindström, K. (2006). Psychosocial approach to occupational health. In G. Salvendy (Ed.), *Handbook of human factors and ergonomics* (pp. 801–817). New York: Wiley.

Landy, F. W. (1985). *The psychology of work behavior* (3rd ed.). Homewood, IL: Dorsey Press.

Lyubomirsky, S., King, L., & Diener, E. (2005). The benefits of frequent positive affect: Does happiness lead to success? *Psychological Bulletin, 131,* 803–855.

Maslach, C. (1982). Understanding burnout: Definitional issues in analyzing a complex phenomenon. In W. S. Paine (Ed.), *Job stress and burnout* (pp. 29–40). Beverly Hills, CA: Sage.

Maslach, C., & Jackson, S. (1984). Patterns of burnout among a national sample of public contact workers. *Journal of Health and Human Resources administration, 7,* 189–212.

Maslach, C., Schaufeli, W. B., & Leiter, M. P. (2001). Job burnout. *Annual Review of Psychology, 52,* 397–422 [S. T. Fiske, D. L. Schacter, & C. Zahn-Waxler (Eds.)].

McMillan, L. H. W., O'Driscoll, M. P. & Burke, R. J. (2003). Workaholism: A review of theory, research, and future directions. In C. L. Cooper & I. T. Robertson (Eds.), *International review of industrial and organizational psychology* (Vol. 18, pp. 167–189). New York: Wiley.

Ministry of Health and Care Services. (2013). White Paper no 29 (2012–2013): *Morgendagens omsorg* [The health care of tomorrow]. Oslo: Ministry of Health and Care Services. Retrieved from https://www.regjeringen.no/nb/dokumenter/meld-st-29-20122013/id723252/.

Moorman, R. H. (1993). The influence of cognitive and affective based job satisfaction measures on the relationship between satisfaction and organizational citizenship behavior. *Human Relations, 6,* 759–776. doi:10.1177/001872679304600604.

Ng, T. W. H., Sorensen, K. L., & Feldman, D. C. (2007). Dimensions, antecedents, and consequences of workaholism: A conceptual integration and extension. *Journal of Organizational Behavior, 28,* 111–136.

Oates, W. (1971). *Confessions of a workaholic: The facts about work addiction*. New York: World.

Peterson, U., Demerouti, E., Bergström, G., Samuelsson, M., Asberg, M., & Nygren, A. (2008). Burnout and physical and mental health among Swedish healthcare workers. *Journal of Advanced Nursing, 62*(1), 84–95. doi:10.1111/j.1365-2648.2007.04580.x.

Reijseger, G., Schaufeli, W. B., Peeters, M. C. W., & Taris, T. W. (2013). Ready, set, GO! From work engagement to job performance. In S. M. Gonçalves & J. G. Neves (Eds.), *Occupational health psychology: From burnout to well-being* (pp. 287–308). Rosemead, US: Scientific & Academic Publishing.

Roe, R. A. (1999). Work performance. A multiple regulation perspective. In: G. Cooper & I. T. Robertson (Eds.), *International review of industrial and organizational psychology* (pp. 231–335). Chichester: Wiley.

Russell, J. A. (1980). A circumplex model of affect. *Journal of Personality and Social Psychology, 39,* 1161–1178.

Russell, J. A. (2003). Core affect and the psychological construction of emotion. *Psychological Review, 110,* 145–172.

Russell, J. A., & Carroll, J. M. (1999). On the bipolarity of positive and negative affect. *Psychological Bulletin, 125,* 3–30.

Salanova, M., Agut, S., & Peiró, J. M. (2005). Linking organizational resources and work engagement to employee performance and customer loyalty: The mediation of service climate. *Journal of Applied Psychology, 90*(6), 1217–1227.

Salanova, M., Llorens, S., Acosta, H., & Torrente, P. (2013). Positive interventions in positive organizations. *Terapia psicológica, 31*(1), 101–113.

Salanova, M., Llorens, S., Cifre, E., & Martínez, I. M. (2012). We need a Hero! Toward a validation of the healthy and resilient organization (HERO) model. *Group and Organization Management, 37*(6), 785–822. doi:10.1177/1059601112470405.

Salanova, M., Martínez, I. M., & Llorens, S. (2014). A more "positive" look at occupational health from positive organizational psychology during crisis times: contributions from the WoNT research team. *Papeles del psicólogo, 35*(1), 22–30.

Schaufeli, W. B., & Bakker, A. B. (2004). Job demands, job resources, and their relationship with burnout and engagement: A multi-sample study. *Journal of Organizational Behavior, 25*(3), 293–315.

Schaufeli, W. B., & Bakker, A. B. (2010). Defining and measuring work engagement: Bringing clarity to the concept. In A. B. Bakker & M. P. Leiter (Red.), *Work engagement; A handbook of essential theory and research* (pp. 10–24). New York: Psychology Press.

Schaufeli, W. B., Salanova, M., González-Romá, V., & Bakker, A. B. (2002). The measurement of engagement and burnout: A two sample confirmatory factor analytic approach. *Journal of Happiness Studies, 3,* 71–92.

Schaufeli, W. B., Taris, T. W. & Bakker, A. B. (2008a). It takes two to tango: Workaholism is working excessively and working compulsively. In R. J. Burke & C. L. Cooper (Eds.), *The long work hours culture. Causes, consequences and choices* (pp. 203–226). Bingley, UK: Emerald.

Schaufeli, W. B., Taris, T. W., & Van Rhenen, W. (2008b). Workaholism, burnout, and work engagement: Three of a kind of three different kinds of employee well-being? *Applied Psychology: An International Review, 57,* 173–203.

Schleicher, D. J., Watt, J. D., & Greguras, G. J. (2004). Reexamining the job satisfaction-performance relationship: The complexity of attitudes. *Journal of Applied Psychology, 89,* 165–177.

Spector, P. E. (1997). *Job satisfaction: Application, assessment, causes, and consequences.* Thousand Oaks, CA: Sage.

Spence, J. T., & Robbins, A. S. (1992). Workaholism: Definition, measurement, and preliminary results. *Journal of Personality Assessment, 58,* 160–178.

Sussman, S. (2012). Workaholism: A review. *Journal of Addiction Research and Theory, 6.* doi:10.4172/2155-6105.S6-001.

Taris, T. W., Schaufeli, W. B., & Shimazu, A. (2010). The push and pull of work: About the difference between workaholism and work engagement. In A. B. Bakker & M. P. Leiter (Eds.), *Work engagement: A handbook of essential theory and research* (pp. 39–53). New York, NY: Psychology Press.

Taris, T., & Schreurs, P. (2009). Well-being and organizational performance: An organizational-level test of the happy-productive worker hypothesis. *Work & Stress, 23*(2), 120–136.

Thompson, E. R., & Phua, F. T. T. (2012). A brief index of affective job satisfaction. *Group & Organization Management, 37*(3), 275–307. doi:10.1177/1059601111434201.

Warr, P. (2007). *Work, happiness, and unhappiness.* Mahwah, NJ: Lawrence Erlbaum.

Wright, T. A., & Cropanzano, R. (2000). The role of organizational behavior in occupational health psychology: A view as we approach the millennium. *Journal of Occupational Health Psychology, 5,* 5–10.

Wright, T. A., & Cropanzano, R. (2004). The role of psychological well-being in job performance: A fresh look at an age-old quest. *Organizational Dynamics, 33,* 338–351.

Wright, T. A., Cropanzano, R., & Bonett, D. G. (2007). The moderating role of employee positive well-being on the relation between job satisfaction and job performance. *Journal of Occupational Health Psychology, 12*(2), 93–104. doi:10.1037/1076-8998.12.2.93.

Wright, T. A., Cropanzano, R., Denney, P. J., & Moline, G. L. (2002). When a happy worker is a productive worker: A preliminary examination of the three models. *Canadian Journal of Behavioural Science, 34,* 146–150.

Wright, T. A., & Staw, B. M. (1999). Affect and favorable work outcomes: Two longitudinal tests of the happy-productive worker thesis. *Journal of Organizational Behavior, 20,* 1–23.Citation ID="CR72">Xanthopoulou, D., Bakker, A. B., Demerouti, E., & Schaufeli, W. B. (2009). Reciprocal relationships between job resources, personal resources and work engagement. *Journal of Vocational Behavior, 74*(3), 235–244.

Xanthopoulou, D., Bakker, A. B., Demerouti, E., & Schaufeli, W. B. (2012). A diary study on the happy worker: How job resources relate to positive emotions and personal resources. *European Journal of Work & Organizational Psychology, 21,* 489–517.

The manufacturer's authorised representative in the EU is Springer Nature Customer Service Centre GmbH, Europaplatz 3, 69115 Heidelberg, Germany. If you have any concerns regarding our products, please contact ProductSafety@springernature.com

Printed and bound by CPI Group (UK) Ltd, Croydon, CR0 4YY
01/10/2026
02225883-0014